Wetland Drainage, Restoration, and Repair

Wetland Drainage, Restoration, and Repair

Thomas R. Biebighauser

THE UNIVERSITY PRESS OF KENTUCKY

 Published in partnership with Eastern
Kentucky PRIDE

Scholarly publisher for the Commonwealth, serving
Bellarmine University, Berea College, Centre College of
Kentucky, Eastern Kentucky University, The Filson Historical
Society, Georgetown College, Kentucky Historical Society,
Kentucky State University, Morehead State University,
Murray State University, Northern Kentucky University,
Transylvania University, University of Kentucky, University
of Louisville, and Western Kentucky University.
All rights reserved.

All photographs by the author unless otherwise noted.
Drawings by Dee Biebighauser. Central Park photo (negative
#57700) is used with permission from the collection of the
New-York Historical Society.

Editorial and Sales Offices: The University Press of Kentucky
663 South Limestone Street, Lexington, Kentucky 40508-4008
www.kentuckypress.com

16 15 14 13 12 11   6 5 4 3 2

Library of Congress Cataloging-in-Publication Data

Biebighauser, Thomas R.
  Wetland drainage, restoration, and repair / Thomas R.
Biebighauser.
      p. cm.
  Includes bibliographical references and index.
  ISBN-13: 978-0-8131-2447-6 (hardcover : alk. paper)
  ISBN-10: 0-8131-2447-6 (hardcover : alk. paper)
  1. Wetland restoration. 2. Drainage. I. Title.
  QH75.B498 2007
  333.91′8—dc22
                              2007001271

This book is printed on acid-free paper meeting the
requirements of the American National Standard for
Permanence in Paper for Printed Library Materials.

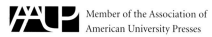 Member of the Association of
American University Presses

Manufactured in China

To my parents, Harlan and Phyllis Biebighauser,
who encouraged me to explore wetlands

Contents

Acknowledgments

I am indebted to the following individuals who spent time with me in the outdoors showing and describing drainage activities known to them: Philip Annis, Richard Bond Jr., Donnie Centers, Dewice Copher, James Duininck, James Flowers, Don Hurst, Harold Korthuis, Jimmy Lyons, George McClure, David Murphy, Richard Neal, John Newman, Billy Osborne, Earl J. Osborne, Cindy Ragland, Rand Ragland, Eddie Ratcliff, Randy Smallwood, John D. Smith, Greg Spencer, Kim Spencer, Merlin Spencer, Ed Stevens, David Taylor, Dave Watson, Scott Whitecross, and Wayne Williams.

I thank these individuals for obtaining and sharing information with me about wetland projects in their area: Phyllis Allison, Ronetta Brown, Tommy Counts, Dennis Eger, Brad Feaster, Denny FitzPatrick, Kathy Flegel, Margaret Golden, Jon Hakala, Richard Hunter, James Kiser, Barb Leuelling, Scott Manley, Beverly McDavid, John Meredith, Doreen Miller, Terry Moyer, Eddie Park, Sherry Phillips, Amy Ross, Thad Ross, Charlie Schafer, Albert Surmont, Paul Tine', Wes Tuttle, Ewell Vice, and Gerald Vice.

The following persons volunteered to provide me with information about wetlands and management activities: Mike Armstrong, Neil Calvanese, Lenny Cecere, Randy Cox, Tammy Cox, Jim Curatolo, Gary Dearborn, Gerry Dilly, Jack Ellis, John Faber, Danny Fraley, Pat Hahs, Lacy Jackson, John Keenon, Eddy Kime, Kathleen Kuna, James Lempke, Doug Malsam, Marty McCleese, Evelyn Morgan, Lisa Morris, Arthur Parola Jr., Wayne Pettit, Merrill Roenke, Ronald Taylor, Anthony Utterback, John Utterback, Sonny Utterback, Jason Vance, Ann Vileisis, Ricky Wells, Charlie Wilkins, Chris Yearick, and Melissa Yearick.

I sincerely thank Lucia Parr and Cynthia Garver for editing the entire book. Their skill, patience, and many suggestions greatly improved the end result. I also thank these individuals for reviewing various chapters in the book: Frank Bodkin, Tony Burnett, Dennis Eger, Julia Ewing, Barbara Leuelling, Arthur Parola Jr., and Christopher Pearl.

Frank Bodkin, Missy Eldridge, Wes Mattox, and Lin Whitley provided regular encouragement and support to me for this project. I owe Frank and Missy much for helping with computer problems, images, and design work that saved me a great deal of time.

I greatly appreciate the frequent help that my wife Dee, daughter Emily, and son Eric gave to me. I also thank Dee for sketching all of the drawings for this book. Without her support and encouragement, this book would not have been possible.

Richard Thomas, Karen Deaton, Cindy Lackey, Jennifer Johnson, and Mike Pollard from Eastern Kentucky PRIDE provided me with frequent encouragement and the support needed to proceed with the publication of this book.

Introduction

Wetlands are the best places to visit if you want to see wildlife. The sight and sounds of startled wood ducks and honking geese rising from the marsh can provide quite a thrill. Helping young people investigate wetlands by netting tadpoles and fairy shrimp is an experience everyone will remember. Communities are discovering the many values of wetlands and are even beginning to build them as tourist attractions. They're also realizing how important these ecosystems are for reducing the severity of floods and for improving water quality.

The purpose of this book is to help people build wetlands. By following the techniques described herein, you can be sure that the wetland you build will hold water. True stories are told about how people who are not wetland scientists have done excellent work creating wetlands that look and function like the real thing. This book will reassure you that it is okay to tear up the ground and make a mess when building a wetland. You'll see that destroying wetlands wasn't pretty, and neither is restoring them a clean, attractive procedure.

In order to succeed at wetland restoration, it is critical to learn how to recognize where actions have been taken before you to disable these ecosystems. I have divided the chapters in this book into two sections. Chapters 1 through 10 describe how wetlands were drained to encourage agricultural production; chapters 11 through 18 provide guidelines on returning the drained lands to wetlands. Since there is a lot of money out there to help build wetlands, I conclude the volume in chapter 18 by providing information about how you can apply for a grant to help pay for wetland restoration.

This book begins by highlighting explicit directions given to landowners beginning in the 1700s, explaining how to successfully drain wetlands. I have dug into a progression of land drainage books in order to report on the many ways used to dry wetlands for agriculture, mining, and development. These printed guides reveal an advanced, organized, and thorough knowledge of wetland genesis and groundwater dynamics that were readily shared with peoples nationwide.

Specific examples show how closely these directions were followed to drain from very small to very large wetlands. By actually watching individuals drain wetlands and talking with them about their techniques, I have confirmed how effective and profitable these actions have been. We then visit the actual locations of wetlands that were drained years ago, often with the same people who drained them. These visits validate that modifications made to landscapes to eliminate surface waters and groundwaters are still working today. Clues are then uncovered that can be used to reveal where wetlands occurred hundreds of years after they were drained and filled.

I admit that for years I didn't think well of people who drained wetlands. Certainly, they were aware of the tremendous value of these unique ecosystems yet chose to destroy them anyway. After all, who would bury such a habitat so full of life? I have now met many individuals who have destroyed wetlands; some have even become my friends. I see now that the drainers of wet land were kind, hardworking, intelligent people who placed a much higher value on fields that could be farmed. These folks were extremely dedicated to their cause of converting what they viewed as worthless swampland to acres of great value to their families. So zealous were they in their efforts that they would toil for years and spend large amounts of their own money to drain and fill these areas.

My research into how wetlands were drained finds a long-term partnership in which individuals and their government joined hands to convert swampland into farmland. The settler's perseverance, hard work, and production have made astonishing changes to what we see on our landscape today. I've been amazed by the many ways people have found to eliminate wetlands. If a diversion ditch did not work, they hauled in fill. If fill did not work, they went to much greater expense by burying drain lines until the land was dry.

The stories in this volume illustrate that the achievement of draining a wetland was not an easy project. It often involved a considerable amount of work and expense over a long period of time. To be successful, the farmers had to be persistent, willing to engage in exhausting physical labor, and spend their limited dollars on heavy equipment and supplies with no guarantee of good results. They truly believed that what they were doing was what was best for their families and for their communities. And due to their willingness to share what they learned, both good and bad, with other landowners experiencing the same problems, we have their records to acquire a deeper understanding as we seek to restore wetlands to their original natural states.

My intent is to encourage restoration projects by showing how men and women are currently helping to restore wetlands. The book describes simple techniques in as much detail as that given to people for draining wetlands in generations past. You will see that just as wetlands were not always drained in the first attempt, they are not always restored the first time, either.

Henry French said in his 1884 book *Farm Drainage* that a knowledge of the depth to which the water table should be removed and the means of removing it constitutes the science of draining.[1] Perhaps we could agree that understanding how an area was changed and the practice of returning a water table to its historic elevation constitutes the art and science of wetland restoration. Draining all these wetlands did not come easy, and neither will their restoration. Returning wetlands to the landscape is rewarding, yet difficult. Working long days and investing personal funds is common when building a wetland. It will take tremendous dedication and much plain hard work to bring back what many consider to be the most beautiful and threatened ecosystems in North America.

When I began writing about wetland restoration, I thought that simply recording a couple of techniques would cover most opportunities. I was wrong. Just as with drainage, there are many ways to construct a wetland. I am pleased to report that people I have talked to about building wetlands are passionate about their cause, and I hope their passion is contagious. These stories are transcribed to inspire you to accept the challenge of wetland restoration and to show how to be successful in your mission.

Throughout this book I use the words "restoration," "creation," and "establishment" to describe wetland projects. *Restoration* explains actions designed to return to a condition that existed about the time of European settlement. *Creation* depicts building a wetland where there are no signs one ever existed, and *establishment* is used when we really don't know if the project is a restoration or a creation. Wherever possible, the various wetland types described in this book are named using the *Classification of Wetlands and Deepwater Habitats of the United States* by the U.S. Fish and Wildlife Service.[2]

CHAPTER ONE

Ages of Drainage

One doesn't have to look far to find a television, newspaper, or magazine story that is expounding on the many values of wetlands to society. Unfortunately, society's appreciation for these productive ecosystems is a rather recent phenomenon. Historically, we find that many who owned wetlands did not share our modern admiration of these ecosystems. To understand why wetlands have been maligned, it is imperative to realize that there has always been great incentive for their removal. To the subsistence or tenant farm family, the acreage gained by draining a marsh could mean the difference between life and death. To the better-off landowner, converting wetlands to farmland translated into improved yields, higher profits, and a greater price at the time of resale. Wetland drainage has even allowed cities and towns to focus development in specific areas instead of spreading it over the countryside.

The first known written guidance regarding conflicts between wetlands and agriculture was given by Cato in the second century:

In the winter it is necessary that the water be let off from the fields. On a declivity it is necessary to have many drains. When the first of the autumn is rainy there is the great danger from water; when it begins to rain the whole of the servants ought to go out with sarcles, and other iron tools, open the drains, turn the water into its channels, and take care of the corn fields, that it flow from them. Wherever the water stagnates amongst the growing corn, or in other parts of the corn fields, or in the ditches, or where there is anything that obstructs its passage, that should be removed, the ditches opened, and the water let away.[1]

For centuries people have taken action to remove water from fields, lest they remain wet land and their planted mainstay be drowned by standing water. Columella, who lived in the time when Augustus and Tiberius ruled the Roman Empire, wrote:

When the soil is moist, ditches are to be dug in order to dry it up and let the water run off. We know of two kinds of ditches: those which are hidden and those which are wide and open; as to the hidden ditches, one will dig out trenches three feet in depth, which shall be half filled with small pebbles of pure gravel, and then the whole will be covered with the earth which was taken out from the trench.[2]

These drainage measures described by Columella were used only in swampy places, or where the soil was saturated with water from springs, for the purpose of protecting crops from surface water; this occurred mainly in the fall and winter.[3]

Directions describing how to drain wetlands for agriculture can be found in an anonymous 1583 publication from England, where it is stated, "Herein is taught, even for the capacity of the meanest, how to drain moores [sic], and all other wet grounds or bogges [sic], and lay them dry forever."[4]

John Johnstone of Edinburgh, Scotland, author of *An Account of the Mode of Draining Land,* wrote:

In 1770, Mr. Poole of that county (Sussex) informed a farming traveler, "that near one hundred years ago, a very large oak, two hundred years old, was cut down at Hook. In digging a ditch through the spot where the

3

old stump was, on taking up the remains of it, a drain was discovered under it, filled with alder branches; and it is remarkable, that the alder was perfectly sound, the greenness of the bark was preserved, and even some leaves were sound. On taking them out, they presently dropped to powder. It is hence very evident, that underground draining was practiced three hundred years ago in this Kingdom.[5]

It is interesting that the underground drains described by Johnstone would have been constructed around 1400 A.D.

In 1764, President George Washington, with five partners and the financing of 100 investors, applied to the Virginia Assembly to obtain a charter for a new company, Adventurers for Draining the Great Dismal Swamp, later called the Dismal Swamp Land Company.[6] The investors were interested in reaping benefits from the timber in the swamp, especially the large bald cypress trees, and later in the agricultural land that would be created after the lumber harvest. Washington's company thereby ordered slaves to dig canals and ditches that drained portions of the Great Dismal Swamp in the late 1700s.[7]

John Johnston of Knock Knolling, Dalrys, and Dumfries Shire, Scotland, was a sheep farmer who moved to America in 1821.[8] He began looking for land to buy in New York State and was offered a farm at a reasonable price in the Rochester area. He decided not to settle in what would later be known as Four Corners, as it was a wet area and a place where malaria was thought to be common. He eventually acquired 320 acres on the shore of Seneca Lake near Geneva in the Finger Lakes region. Geneva appealed to him because a college was being built there, and the character and intelligence of the people impressed him. Agnes Hutchins, John Johnston's granddaughter, wrote: "After a few years of farming in America, my grandfather found that twenty bushels of wheat to the acre was all that it was possible to raise because water would stand on the clay soil and the young wheat would die out in these spots in the spring."[9] In 1838, he began draining his farm by burying clay tiles in the ground, and eventually he produced

60 bushels of wheat per acre from the same wet fields. Johnston thereby introduced, pioneered, and ardently promoted the use of clay tiles to drain lands for agriculture in North America.[10]

Johnston recommended that lands with saturated soils be drained to increase productivity, suggesting the following test to determine if an area needed draining: "Dig holes about two and a half feet deep in different parts of the field; put a cover on the holes so that rainwater cannot get into them, and if they fill with water until within a foot or so of the surface, in ten or twelve hours, then his land requires, and will pay well for draining." He went on to say: "To the unpractised eye, land that looks dry is gorged with water six inches below the surface." He described how lands with wet soils and a high water table, which we now know are responsible for the maintenance of ephemeral wetlands, wet meadows, and forested wetlands, were improved by draining.[11]

Over the course of his lifetime, Johnston had 72 miles of clay drain tiles installed on his 320-acre farm.[12] Stories of his success with draining spread far and wide: "The beneficial effects of draining on Mr. Johnston's farm are very apparent. Places which formerly would bear no wheat, nor indeed scarcely anything but a kind of sour grass and reeds, are made, merely by draining, to produce the finest crops of every description of grain."[13] Johnston was well respected for his contributions to agriculture in North America and became known as "the father of tile drainage."

Early leaders who recommended wetland drainage in the United States paid close attention to the techniques being developed for converting fens, moors, and bogs to cropland in Europe. In 1797, Johnstone wrote the first edition of his detailed book about land draining in England. He stated that wetlands were formed by two main causes: by rainwater that was kept standing on the surface in clay-lined depressions, and by water that emerged from springs confined in the ground.[14] Henry French, who farmed in New Hampshire, reported visiting England in 1857 for the purpose of learning about drainage practices before publishing the first edition of his widely read book *Farm Drainage* in 1884.[15] Frederick Law Olmsted, considered America's first and most renowned landscape architect, traveled to Great Britain, where he learned of the value of deeply buried clay tile drainage systems from celebrated drainage engineer Josiah Parks and later used this knowledge to help design the wetland drainage system for Central Park.[16]

It may be hard to believe that Central Park in New York was largely wetland at one time. Drainage author Charles Elliott wrote that before improvement, the 856-acre area was "regarded as a menace to the health of the city."[17] Colonel George Waring used English methods in 1858 to drain Central Park in what was considered to be the largest drainage project for its time. In his book *Draining for Profit, and Draining for Health*, Waring provides particulars on a portion of the huge effort: "The tract drained by this system, though very swampy, before being drained, is now dry enough to walk upon, almost immediately after a storm, except when underlaid by a stratum of frozen ground."[18] To accomplish the drainage, workers hand-buried over 60 miles of clay pipes, 1.5 to 6 inches in diameter, in ditches usually 4 feet deep along parallel lines spaced at 40-foot intervals. Waring's effort required precise application of civil engineering principles as a number of the large drain lines were positioned to have only 1-inch drop for every 1,000 feet.[19]

Henry French, another drainage writer who respected George Waring's work, said that the Central Park project marked the first time round drain pipes with collars were used in the United States.[20] These round pipes were regarded in England to be superior to the horseshoe type used by John Johnston for earlier projects.

In combination with the buried drain lines, an enormous amount of soil was brought in to Central Park to fill the low lands: 10 million one-horse cartloads carried 5 million cubic yards of stone and earth to the site from Long Island and the New Jersey meadowlands, a feat equal to a single procession of 30,000 miles. The moving of the materials alone was enough to change the elevation of the park by roughly 4 feet.[21]

Hand-dug ditch hollowed out in the 1700s by slaves owned by the Dismal Swamp Land Company, of which George Washington was a partner. The ditch was used to drain part of the Great Dismal Swamp in Suffolk County, Virginia. (Ann Vileisis photograph)

Historian Barry Lewis says that the drainage system Vaux designed and Waring installed "still allows the above ground fantasy we see today" at Central Park. He further explains how some of the marshes in Central Park were excavated by Waring to form ponds and lakes.[22]

Over the years, when completing reconstruction projects in Central Park, Senior Landscape Architect Gary Dearborn, Senior Vice President Chris Nolan, and Vice President for Operations Neil Calvanese have encountered portions of the drainage system installed by Waring. Gary says that he has "been into some of his pipes"—namely, the larger 18-, 24-, and 36-inch drains constructed by Waring—and that he often ties new drainage structures into these early buried structures of brick masonry and stone.

Gary Dearborn describes how they unearthed a number of the smaller-diameter clay drain tiles in the 1980s while significant portions of the park were being rebuilt. The clay tiles found are basically round and appear to have been formed by extrusion from a mold, and then joined with collars as described by Henry French. Some of the buried clay tile lines were still carrying water, and the ground surface above them remained dry. Ac-

Emergent wetland in 1857, the year before it was deepened by Colonel George Waring to form a lake for drainage water in Central Park. This area is now called the 59th Street Pond. (From the Collection of The New-York Historical Society)

cording to Gary, in 1992, Chris Nolan dug into a system that had been buried by Waring along the base of Great Hill, apparently for trapping a spring. Water gushed from the hole that had been excavated after Nolan's group hit the clay tile line. Gary says that some of the larger fields in the park drained by Waring have since been converted to ball fields. He believes that portions of the extensive system of smaller-diameter clay pipes installed by Waring continue to work today in those areas that had not been affected by extensive reconstruction efforts that took place in the 1930s and 1980s.[23]

Neil Calvanese has watched contractors dig up Waring's original clay drainage tiles on a number of occasions. He says that some of the lines were carrying water, some were dry, and others were plugged with soil. When asked about soil colors near the buried lines, he says they varied from brown to gray. Neil has found oval-shaped tiles, flat-bottom tiles, and round tiles joined with loose-fitting collars. He believes that most of the

outlets for the original drainage system have been blocked by the many reconstruction actions in the park.[24]

An enlightening report on how wetlands were drained for farming was given by Dr. Gray in 1865 concerning an asylum in Utica, New York:

We commenced the work of under draining fifteen years ago, upon a garden containing twenty-two acres, which was a swamp, and eight acres of it had never been plowed. By under draining, the land was put in condition to yield large crops. We consulted Mr. Thomas, and made a trial of tile and wood. The tile is expensive, and difficult to arrange at the joints. The sand gets in and chokes the drain. We finally adopted wooden drains, as cheaper and easier to put down. We take a hemlock plank, eleven inches wide, slit in two pieces, and nail together; then lay a twelve inch board on the floor or bottom of the ditch, and upon this put the V shaped trough, breaking joints at the connections, by nailing boards at these points so as to be tight. Small stones are thrown in on the sides and top, and then covered. The whole cost was fifty cents per rod, when put in by hired labor. Much of the work was performed by inmates of the asylum.[25]

On April 19, 1870, a discussion was held at a meeting of the Farmer's Club in New York City:

Mr. Thomas Johnson Perry, Ohio. I have a farm of 120 acres, formerly covered with timber, and much of it under water half the year. I have cut off the timber with my own hands. I have also made 1,200 rods of open ditches, in depth from two to three and one-half feet, and from eight to twelve feet wide at the top. I use a plow and scraper, taking the earth into the field, filling up all the low places. I have under drained with brush; that did no good—the crabs[26] shut it up. I then tried timber, laying a rail at each side of my ditch, then covering with oak-heading two feet long and two inches thick, laid crosswise. This answers a very good purpose as long as it will last—about twenty years in drains that are dry half the time; when there is water all the time it would last longer. I am now using tile from two to five inches inside of the pipe. I am putting them down from two to five feet deep. At the bottom of my drain I cut a groove the size of my tile. I then commence at the upper end of my drain to lay tile, pressing them down into the groove. I choose a time when there is water in the ground. I put a good hard brick at one end of the first tile, pressing each one down until the water will pass through, clearing out all loose earth that may fall in with a tool made for that purpose. My experience is that land that is not worth five dollars per acre for farming purposes without being drained, is worth sixty dollars after having been thoroughly drained. It will cost about twenty dollars per acre to drain land here. We pay $1.25 per 100 feet for two-inch tile, $2.10 for three-inch, three dollars for four-inch, four dollars for five-inch, $4.50 for six inch. I have raised 120 bushels (ears) of good sound corn per acre, twenty-seven bushels of wheat, sixty bushels of oats, and this on land which would not do anything without drainage.[27]

John Klippart, who first published a book on land drainage in 1861, said, "Wet and swampy lands, when thoroughly drained, are found to be among the most productive, and hence their im-

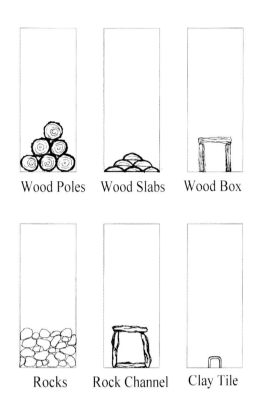

Wood Poles   Wood Slabs   Wood Box

Rocks   Rock Channel   Clay Tile

Early types of buried drainage structures placed in hand-dug ditches. (Drawing by Dee Biebighauser)

provement by drainage is most marked and satisfactory."[28] The installation of drain tiles was an effective way to eliminate wet meadows, spring seeps, and ephemeral wetlands, as evidenced by a statement made by R. J. Swan of Seneca County, New York:

In the year 1851, I laid 16,000 tiles, which with the quantity laid this year, completes the drainage of 200 acres of my farm. I deem it to be thorough, and it is so esteemed by others who have preceded me in this essential system of farm management. One important feature, discernable, is the reclaiming twenty-four acres of soil which had never been tilled, producing only coarse aquatic grasses unfit for hay or pasture; this portion of the farm has hitherto been charged with the interest of cost and also taxes, without any return, a portion which hereafter will yield crops of any grain or grasses, equal to any other land.[29]

An early view toward wetlands is found in the appendix to *Tile Drainage* by W. I. Chamberlain,

Emergent wetland near Clearfork Creek in the Daniel Boone National Forest in Kentucky. This type was once common in mountainous areas. Fragrant white water lily, a rare plant for the area, began growing naturally in this wetland after its construction in 1996.

where A. I. Root, publisher as well as author of the appendix, writes:

In riding over the country, whenever I see water standing in cat-swamps[30] or sink-holes, doing nobody any good, and damaging crop, or the chances of crop, I feel a strong impulse to let the water off. If the owners of such places enjoyed the work as I do, I verily believe they would sit up nights to drain off these eyesores on the land, if they could not manage it otherwise. How I do like to see the water run away, like a liberated bird! And then to witness the dismay of the frogs, turtles, and other denizens of such places, is worth almost as much, if not quite, as to see the wonderful crops which always reward such labor.[31]

Henry French wrote in his book *Farm Drainage* that, upon hearing the federal government planned to turn over close to 60 million acres of swamp and overflowed lands to the states, Governor Wright of Indiana gave an address in which he estimated the marshy lands in his state to be at 3 million acres. The governor said: "These lands were generally avoided by early settlers, as being comparatively worthless; but, when drained, they become eminently fertile. I know a farm of 160 acres, which was sold five years ago for $500, that by an expenditure of less than $200, in draining and ditching, has been improved, that the owner has refused for it an offer of $3,000."[32]

Efforts to drain wetlands for agriculture in the United States became well organized and funded in the late 1800s. In the third edition of *Engineering for Land Drainage*, Charles Elliott outlined a systematic approach for draining both large and small acreages to provide for the "profitable production of crops." Chapter 16 in his book is devoted to the establishment of drainage districts, defined as

an organization of the owners of land formed for the purpose of constructing and maintaining adequate drainage outlets whose cost shall be shared in proportion to the benefits derived. The kinds of lands properly subject to such organization are swamps or wet lands, wholly or partially unreclaimed; farm lands which have insufficient outlets; lands in river and creek valleys which are subject to overflow; and coast and tidal lands subject to inundation by the sea.[33]

Elliott told how draining a large area or an entire watershed is accomplished in two parts: publicly funded drainage to produce necessary outlets for water, and privately funded drainage on individual farms to direct water into these constructed outlets. Elliott served as chief of drainage for the U.S. Department of Agriculture (USDA), and his knowledge of drainage was so valued that the first and second printings of his book sold over 13,000 copies, requiring that a third edition be written and printed.

Using covered drain tile ditches to create and improve farmland became such an accepted practice that more than 75,000 miles of clay tile were placed in New York State before 1900.[34] Weaver reported that during 1850–1859, 66 clay tile factories were operating in the United States, increasing to 234 in the next decade, and up to 840 by 1879.[35] By 1882, there were 1,140 factories manufacturing clay tiles in the United States.[36]

There was a genuine concern that the presence of wetlands contributed to poor health in a community, as printed by Henry French: "Frogs and snakes find in these swamps an agreeable residence, and wild beasts a safe retreat from their common foe. Notoriously, such lands are unhealthful, producing fevers and agues in their neighborhood, often traceable to tracts no larger than a few acres."[37]

The draining of wetlands did not always end in success. An 8- to 10-acre field in Georgia, described as an old cypress pond, was taken out of cultivation thirty years after drainage when the men who worked it experienced tremendous itching, and their mules' feet became sore and inflamed.[38] Residents first blamed the sickness on hook worm, pollen, alkali, and even sulfuric acid in the soil.[39] An investigation by state geologist S. W. McCallie found that the illness was caused by the spicules of freshwater sponges left in the soil.[40] These freshwater sponges are often found in wetlands, lakes, and streams in North America. Needlelike spicules of silica and calcium provide the skeleton for their soft body tissue. When the sponges died, their spicules remained in the soil, and the sharp spines cut flesh like slivers of glass. Davis recommended completing a microscopic examination of soils from a lake bed or swamp where drainage was contemplated, and if spicules were found that sand be added and the top 6 inches of soil be mixed with deeper soils to lessen unwanted effects.[41] Professor John Haswell speculated that sponges may also have been responsible for the itching reported by other individuals in other locations who worked on muck soils and peat lands. To prevent this problem from developing in the future, he concluded that the study of areas before drainage could easily include an examination for spicules.[42] I have found no evidence that his advice was ever heeded.

John Keenon remembers when most of the corn raised in Powell County, Kentucky, was up on the higher ridges and benches overlooking the Red River outside of Stanton, formerly known as Beaver Pond. The level bottomland soils along the river had a hardpan, with abundant burrowing crayfish, and were too wet to farm. His father,

Aerial view of the Brewer Farm in Powell County, Kentucky, along the Red River immediately after the installation of a drainage system that was designed and funded by the Soil Conservation Service in the late 1960s. A backhoe was used to bury over 70,000 feet of drain lines in a grid-like pattern, passing under roads and down constructed waterways over a considerable distance to reach the Red River. Before this drainage, the lands were too wet to crop. (Photograph taken by an employee of the USDA Natural Resources Conservation Service who assisted with the project)

owning over 4,000 acres, hired dozer operators to clear hundreds of acres of bottomland hardwoods, including sweet gum and willows along the river. To drain these swamps, throughout the 1950s he hauled in truckloads of clay drain tiles from Indiana and placed them in miles of hand-dug ditches.[43] This massive drainage program was completed with the assistance of the USDA Soil Conservation Service (SCS) and greatly changed the face of agriculture in the mountains of eastern Kentucky by making it possible to raise corn and quality hay along the Red River. Don Hurst, USDA Natural Resources Conservation Service (NRCS) district conservationist for the area, claims that a majority of the thousands of acres of fields along the Red River have been drained with tiles.[44] Keenon says that the clay tiles his father installed continue to work to this day and that he still maintains the outlets.

When the 1985 Farm Bill passed in the United States, it contained a provision for protecting wetlands that eventually became known as "Swampbuster." Landowners would be denied federal commodity price supports, disaster payments, insurance, and loans for activities that would violate its conditions.[45] As a result of its passage, the

SCS was directed to end its massive cost-share program of drainage activities on private land. Randy Smallwood, district conservationist with the NRCS in Bath County, Kentucky, said that Swampbuster came as a large shock to SCS employees and farmers who had spent their lives working to make fields more productive.[46]

It would be naive to believe that the new direction contained in Swampbuster would be readily embraced by field personnel whose careers had been dedicated to helping farmers drain lands for crop production. As landowners applied for benefits under the new bill, the SCS reviewed the fields involved to identify hydric soils, hydric soil inclusions, past crop activity, and evidence of drainage. Upon finding these signs, the fields were labeled "Prior Conversion," or "PC," to represent wetlands destroyed or affected before passage of Swampbuster. Private owners were allowed to maintain drainage features such as ditches and buried drain lines on areas labeled PC, and they did so with continued technical assistance from the NRCS. Since just about every wetland located on agricultural land contains some evidence of past drainage action, most existing wetlands were labeled PC and remained legal for further drainage. The final consequence of Swampbuster was that the role of NRCS was modified to one in which it would now recommend, design, and facilitate, but not directly finance, drainage projects.

The duties of the NRCS in wetland management changed greatly again with the passage of the Farm Bill in 1996. The agency was given re-sponsibility for implementing the Wildlife Habitat Incentives Program (WHIP) on private lands. WHIP was similar to other conservation programs administered by NRCS over the years; however, it provided technical and financial assistance to landowners and others to develop upland, wetland, riparian, and aquatic habitat areas on their property.[47]

Soon after purchasing our 14-acre farm in the mountains of eastern Kentucky, I began working with NRCS District Conservationist Marty McCleese from Rowan County to identify practices for which I would be eligible under the WHIP program. We outlined a variety of actions for my farm, including planting prairie grasses, erecting bluebird and waterfowl nest boxes, installing bat roosting boxes, and, notably, establishing wetlands. Two months later I received a call from Marty that my application (the only one submitted from Rowan County) had been approved, and the next step in the process would be to stop by his office to sign the 5-year contract. Walking into the NRCS office, I introduced myself to the receptionist. While I was waiting for Marty, a number of employees gathered around the counter to talk with me. They said that they just had to meet the person who wanted to take perfectly good bottomland and turn it into a swamp. We all laughed about my decision, but their reaction shows the magnitude of attitude change required of public servants who have spent their entire careers draining lands and are now being directed to promote their restoration.

CHAPTER TWO Why They Pulled the Plug

Farmers have fought water and wetlands to both create and improve areas for crop production over many centuries. Thorough drainage has been considered the first and most important step to profitable crop growing on a large proportion of the farms in the country.[1] Undoubtedly, some immigrants brought with them views of wet ground similar to those recorded by Sir Charles Coote in a survey of agricultural land in King's County, Ireland: "This country abounds with a fine rich grass, and only wants draining and gravelling to be made of the best quality . . . but the fields are greatly overrun with rushes, which could easily be destroyed by draining, and successive cutting, for two or three years."[2]

Many immigrants who settled Appalachia were subsistence farmers, generally owning such small tracts of land that their families' survival hinged on the additional crops that could be produced from a drained wetland. These families were fortunate to own 5 acres of tillable land, and their situation in life could be greatly improved by making a 1-acre beaver pond grow corn. Land that was flat enough to grow crops was at a premium, and a family could greatly increase its financial standing by creating even a small tobacco field from a flooded area.

Farmers have tried for years to grow crops in wetlands with little or no success. Even if farmers wait until summer weather has dried the surface enough to plow and plant, the soils in a seasonal wetland are still likely to clump, requiring repeated passes with a plow to break up the clods. John P. Norton, professor of agricultural chemistry at Yale University, made the following comments about attempting to farm wet areas in 1848: "But the evil effect of such water upon the soil, is seen, not in bogs and swamps alone, but also in a great number of our cultivated fields. In such places, water is not present to the extent before described; the soil may be even perfectly firm and dry in midsummer, but still there is so much water during autumn and spring that neither grass nor cultivated crops succeed well." Norton went on to provide tips on how to identify wet areas that should be drained:

A practiced eye will soon detect these wet fields, or the wet spots caused by concealed springs on land otherwise dry. A few rushes, or some coarse, wiry grass, will always betray the secret. Here too, the only remedy lies in the drain. Its ameliorating influence is more quickly felt on this cold, sour land than in swamps, because the evil has not proceeded so far. I am scarcely acquainted with a farm in my own part of the country which has not some land upon it that needs draining. In nearly every section of New England, I believe that a farm without some wet places on it would be an exception to the rule.[3]

Wayne Pettit began plowing fields with a pair of mules on his father's farm along the North Fork of the Licking River near Bangor in Rowan County, Kentucky, when he was only 13 years old. He used a turning plow to prepare fertile bottomland soils for corn and tobacco planting in the 1950s. Wayne knows what it's like to get mules stuck in the blue-black soils of a wetland:

Before you know what's going on it happens, it marshes up just like quicksand and they can get stuck up to their belly. The mules fight it some before you can help them. To get them out you first unhook the reins and then the plow. Most of the times they lunge around and get up themselves up, but sometimes you have to get someone to help you put a rope around them and pull them out.

Wayne's father taught him how to tell when it was dry enough to plow: "He told me to take a handful of dirt and make a ball. If it stuck together it was too wet to plow, if it fell apart it was dry enough to work." I asked Wayne what would happen if he went ahead and plowed wet ground. He explained that "it ruins the ground because it won't work up. You get nothing but clods the size of your fist that you won't cultivate. Wet ground bakes and gets hard. Corn and tobacco turn yellow and won't do any good." Wayne said he would not plow where water stood because the crops that were planted would be ruined anyway.[4]

Interest in drainage increased as farmers switched from using horses to tractors. Many decided to drain when they discovered that heavy machinery was much more likely to become stuck in areas of saturated soils. Individuals also found that drainage "squared up" a field, making it more efficient to manage.[5]

Most every farmer has a story about getting mired down with a tractor while attempting to plant wet soils. It takes a tremendous effort to free equipment from the grip of a wetland, often requiring a neighbor's even larger tractor, lengths of heavy chain for pulling, and the backbreaking work of shoveling mud. Becoming stuck ruins productivity for the day and, once free of the mud, requires extensive dismantling and cleaning to prevent soil from ruining expensive parts such as bearings. It is not difficult to understand why farmers took measures to dry out these soggy places we call wetlands.

Draining wetlands to create more productive farms was commonplace and became a passion for many. There were clearly recognized benefits for removing standing water and saturated soils from an area with the following real, as well as perceived, benefits credited to drainage:

Draining a swamp, slough, marsh, or other area of wet land can create a productive farm field and improve the size and shape of an existing field.[6]

Drainage removes standing water from snowmelt and from spring and summer storms. Winter wheat is especially susceptible to drowning in the spring, while corn, soybeans, and hay can quickly drown in the summer months.[7]

Wet soils cannot be plowed early in the spring, thereby causing planting to be delayed or even canceled for the year.[8]

Drainage allows for consistent crop production year after year, and fields can be plowed even in an exceptionally wet spring.[9]

Saturated soils take longer to warm in the spring, delaying germination or even causing seeds to rot in the ground.[10]

The roots of planted crops cannot survive in water, and drainage removes this evil.[11]

Drainage produces a longer growing season that is important for crops such as corn in northern climates. It also allows for a greater length of time that fields can be worked.[12]

Drained soils allow crops to mature earlier, thereby avoiding problems with an early frost, or disease, and insect infestations that occur in late summer.[13]

Drained lands produce greater and more consistent crop yields, even in drought years.[14]

Drainage helps keep nutrients contained in applied manure and fertilizer from being carried off during heavy rains and helps move these nutrients into the soil where they can be used by crops.[15]

Drainage of swamps, sloughs, and marshes increases the resale value of a farm.[16]

Drainage improves human health in an area by diminishing malarial, diphtheria, and typhoid tendencies. Animal health is also improved by the elimination of wet soils that can cause black-tongue and rot hooves.[17]

Drainage reduces problems associated with frost moving wheat and clover out of soil.[18]

Tile drainage diminishes runoff, reducing erosion caused by the suddenness and violence
of floods.[19]

Tile drainage removes excessive wetness that kills
timothy and clover, and reduces the presence
of weeds that tolerate flooding.[20]

George McClure from Wheelersburg, Ohio,
has operated a drainage business since 1972, completing jobs in three Kentucky and eight Ohio
counties. He finds that installing buried drain
lines extends the amount of time a farmer can operate equipment in the field by a month or more.
This can make a critical difference when it comes
to picking corn, as heavy rains often saturate soils
at about harvest time, making it impossible to operate rubber-tired equipment in the fields.[21]

Detailed review of agricultural drainage books
written from the late 1700s to the mid-1900s un-
covers uniform guidance on when drainage activities are needed and under what conditions
they prove cost effective to the farmer. What was
clear to all was that when water stood on the surface of the land—as found in a swamp, marsh, or
morass—drainage was needed to grow any kind
of crop.[22] Charles Elliott said: "Ponds and sloughs
are wholly unfit for cultivation, even in the driest
years without drainage."[23]

Following the obvious need to remove surface water, drainage authors move on to make
a strong case for lowering groundwater in fields
by using buried drainage structures. Manly Miles
explained that if the roots of planted crops grew
into the water table, or if the water table rose and
submerged crop roots, those plants would become unhealthy and suffer.[24] Numerous examples
are given on how saturated soils were responsible
for poor yields, inconsistent harvests, and crop
failures.

Miles and Weaver proposed a simple test to
determine when drainage was needed. The trial
involved digging a trench or a couple of holes in a
field, sheltering the holes from rain, then returning the next day to see if water had seeped into the
hole.[25] The presence of water standing in the hole

Farmer's tractor claimed by a wet field during an attempt to plow near Salt Lick
Creek in Bath County, Kentucky.

showed that further drainage was needed. This
type of test is remarkably similar to how we identify the presence of hydric soils in a wetland.

John Klippart in 1861 suggested another way
to identify when drainage was warranted by looking at plants growing on an area; he explained how
lands growing trees such as beech, maple, ash,
elm, or any other kind of timber or shrubs that
required wet ground were seldom tillable, and
certainly never at a profit, until they were drained
with buried structures. He wrote: "Annexed is a
list of plants whose presence is always an unmis-
takable evidence of the necessity of drainage—
because they flourish only in very moist or wet
soil. As soon as the soil is properly underdrained,
all the plants named in the list will disappear, because their accustomed supply of moisture will
then be withdrawn, and they of course, will per-
ish."[26] Table 1.1 is the list he included.

Henry French, in his book *Farm Drainage,* also
mentions the presence of aquatic plants as indi-
cators of too much moisture: "If the land be in
grass, we find that aquatic plants, like rushes or
water grasses, spring up with seeds we have sown,
and, in a few years, have possession of the field,
and we are soon compelled to plow up the sod,
and lay it again to grass."[27]

Farmers were advised to drain even the small, wet places in their fields before they became a bigger problem.[28] If the wet soils were crossed by heavy machinery, compaction might occur that would ruin soil structure. Dry soils often surrounded these pockets of wet soil, making it difficult to avoid them in working the dry areas. The compacted soils would hold standing water longer and enlarge over time, further reducing crop production until they were drained.

Even though the majority of drainage has been done to grow crops, a considerable acreage of wetlands was also drained to facilitate the cutting and growing of trees in the south. Many wetlands were successfully converted to pine plantations by simply draining with ditches. E. A. Schlaudt explained how, in 1950, the drainage of pine forests was a new practice in the Southeast, but after only a few years plans were made to drain about 750,000 acres of wet forests. Much of this drainage was being done by commercial timber and pulp companies responding to the increased value of pine logs.[29]

The fact that the seeds from pine trees would not sprout and grow in water was obvious to those who tried to grow them. In south Georgia, farmers found that, within 2 years of draining cypress ponds, the drained areas began growing pine trees when seed trees were left around the edges.[30] In North Carolina, a coastal plain swamp that grew only bald cypress trees began supporting a dense growth of slash pine from neighboring trees when they were ditched.[31] High water levels in pocosin wetlands were also controlled to convert pond pine (*Pinus serotina*) to more valuable loblolly pine (*Pinus taeda*).[32]

Forested wetlands were drained to make it possible to harvest timber without delays from rain and as a means for increasing profits. Drainage allowed loggers to access trees for cutting and removal without purchasing the costly specialized equipment needed to work in soft ground and standing water.[33]

Companies had an added bonus from draining their wetlands when they found that haul roads could be constructed on the spoil that is generated from excavating primary and second-

Table 1.1

John Klippart's List of Plants Needing Drainage, 1861

Botanical Name	Common Name
Ranunculus alismaefolius	Water plantain or Spearwort
R. sceleratus	Cursed crowfoot
R. pennsylvanicus	Bristly crowfoot
Caltha palustris	Marsh marigold
Nasturtium officinale	Water cress
N. palustre	Marsh cress
Cardamine pratensis	Cuckoo flower
Impatiens pallida	Pale touch-me-not
I. fulva	Spotted touch-me-not
Floerkea proserpinacoides	False mermaid
Rhus venenata	Poison sumac or Dogwood
Sanguisorba canadensis	Canadian burnet
Geum strictum	Avens
G. rivale	Purple avens
Rosa carolina	Swamp rose
Rhexia virginica	Deer grass
Lythroughm alatum	Loosestrife
Nesaea verticillata	[No common name given]
Epilobium coloratum	Willow herb
Ludwigia palustris	Water purslane
Penthorum sedoides	Ditch stone crop
Saxifraga pennsylvanica	Swamp saxifrage
Heracleum lanatum	Cow parsnip
Archemora rigida	Cow bane
Cicuta maculata	Water hemlock
C. bulbifera	Hemlock
Conium maculatum	Poison hemlock
Cornus sericea	Silky cornel
C. stolonifera	Red osier dogwood
C. stricta	Stiff cornel
Cephalanthus occidentalis	Button bush
Solidago ohioensis	Golden rod
S. riddellii	Golden rod
S. patula	Golden rod
S. lanceolata	Golden rod
Helianthus giganteus	Sunflower
Coreopsis trichosperma	Tick seed sunflower
Bidens cernua	Burr marigold

ary drainage channels. It was recommended that roads constructed from organic spoil be capped with a thick surface of sandy clay or similar soil to prevent the material from decomposing, thereby reducing settling and lengthening the life of the road. An added bonus for the timber owners was that the cleared zone, consisting of the main

Table 1.1 continued

Botanical Name	Common Name	Botanical Name	Common Name
B. chrysanthemoides	Burr marigold	*Sagittaria variabilis*	Arrow-head
Helenium autumnale	Sneezeweed	*Platanthera peramoena*	Great purple orchis
Cacalia tuberosa	Tuberous Indian plantain	*Spiranthes latifolia*	Ladies' tresses
		Cypripedium spectabile	Ladies' slipper
Cirsium muticum	Swamp thistle	*Iris virginica*	Blue flag
Lobelia cardinalis	Cardinal flower	*Sisyrinchium bermudiana*	Blue-eyed grass
L. syphilitica	Great lobelia	*Scilla fraserii*	Squill or White hyacinth
L. kalmii	[No common name given]	*Lilium canadense*	Wild yellow lily
Plantago major	Rib grass	*Melanthium virginicum*	[No common name given]
Lysimachia ciliata	Loosestrife	*Veratrum viride*	False hellebore
L. radicans	Loosestrife	*Juncus effusus*	Bog rush
L. lanceolata	Loosestrife	*J. scirpoides*	[No common name given]
Chelone glabra	Snakehead	*J. militaris*	[No common name given]
Mimulus ringens	Monkey flower	*J. tenuis*	[No common name given]
M. alatus	Monkey flower	*Cyperus diandrus*	Galingale
Veronica anagallis	Water speedwell	*C. strigosus*	[No common name given]
V. americana	Brooklime	*Eleocharis obtusa*	Spike rush
V. scutellata	Marsh speedwell	*E. palustris*	[No common name given]
Gerardia purpurea	[No common name given]	*E. tenuis*	[No common name given]
Pedicularis canadensis	Lousewort	*E. compressa*	[No common name given]
P. lanceolata	[No common name given]	*Scirpus sylvaticus*	Club rush
Dianthera americana	Water willow	*S. lineatus*	[No common name given]
Lippia lanceolata	Fog fruit	*S. eriophorum*	Wool grass
Physostegia virginiana	False dragon head	*Eriophorum polystachyon*	Cotton grass
Scutellaria lateriflora	Skullcap	Almost all sedges	[No common name given]
Myosotis palustris	Forget-me-not	*Leersia oryzoides*	White grass
Asclepias incarnata	[No common name given]	*Leersia virginica*	[No common name given]
Polygonum amphibium	Knotweed	*Alopecurus aristulatus*	Wild water-foxtail
P. pennsylvanicum	[No common name given]	*Cinna arundinacea*	Wood-reed grass
P. hydropiper	[No common name given]	*Calamagrostis canadensis*	Blue-joint grass
P. acre	[No common name given]	*Spartina cynosuroides*	Freshwater cord grass
P. hydropiperoides	[No common name given]	*Glyceria elongata*	Manna grass
Rumex verticillatus	Swamp dock	*G. nervata*	[No common name given]
R. conglomeratus	Green dock	*G. fluitans*	[No common name given]
Quercus aquatica	Swamp oak	*Phragmites communis*	Reed
Q. palustris	Water oak	*Holcus lanatus*	Meadow soft grass
Symplocarpus foetidus	Skunk cabbage	*Hierochloa borealis*	Vanilla
Acorus calamus	Sweet flag or Calamus	*Phalaris arundinacea*	Reed canary grass
Typha latifolia	Cat-tail flag	*Milium effusum*	Millet grass
Triglochin palustre	Arrow grass	*Sorghum nutans*	Indian grass
Alisma plantago	Water plantain		

John H. Klippart, *The Principles and Practice of Land Drainage* (Cincinnati: Robert Clarke, 1861), 183–186

ditch, roadbed, and road itself, served as an effective firebreak.[34]

In agricultural land, closely spaced deep ditches were often needed to remove both surface water and groundwater; in forested areas, however, the main concern was surface waters, so widely spaced shallower ditches could be used.

These shallow ditches were spaced approximately 0.25 mile apart in wet areas but could be spaced up to 0.5 mile apart in seasonally flooded locations.[35] Trees would further drain soils in these wet areas once they were established.[36]

Disk plows devised for creating firebreaks and fire plows pulled by tractors were used to

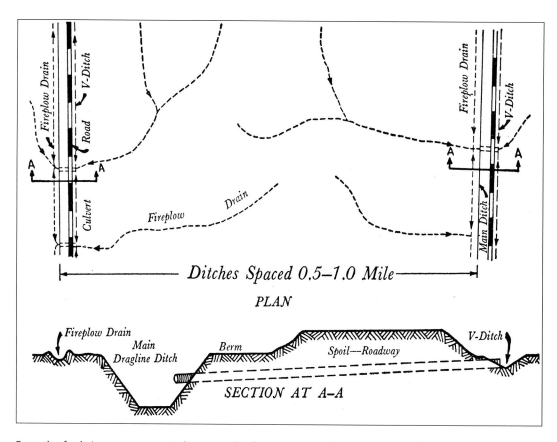

Example of a drainage system created in part with a fire plow in "very flat, wet woodland." (From E. A. Schlaudt, "Drainage in Forestry Management in the South," in *Water: USDA Yearbook of Agriculture* [Washington, D.C.: U.S. Government Printing Office, 1955])

construct small lateral ditches needed to remove surface waters in order to establish trees. Ditches constructed with a fire plow could be designed to meander through the forest, requiring a minimum of clearing and not blocking access to planted areas.[37]

Discovering how the fire plow has been used to drain wetlands intrigued me, as I have fought fires in many areas where they were used to construct fire lines. In my travels around the South as a wild land firefighter, I observed that almost all government forest management headquarters and every timber company office had a dozer and fire plow parked out back for fire suppression. The availability of this equipment, combined with the need to operate it for practice outside of the short firefighting season, has undoubtedly provided opportunities for forest managers to drain wetlands for tree planting and timber harvest.

The ditches created by a fire plow can be expected to last for years. I used a fire plow to test for archaeological resources in a number of different forested areas in 1991 and 1992. The shallow ditches they left behind continue to drain away surface waters 15 years later.

Today, a resurgence in land drainage activities is spreading across the nation with the advent of computerized yield monitors and global positioning systems on harvest machines. New generations of farmers are discovering a strong correlation between crop production and land drainage. Many are finding that it pays, even without cost-share assistance from the government. Businesses carrying drainage products report brisk sales as farmers confirm that well-drained fields produce more crops.[38]

Wetland Drainage, Restoration, and Repair

CHAPTER THREE

Ditching for Dollars

A number of practices have been used and perfected over the years to remove what was considered to be excess water from a tract of land. These techniques generally involve the construction of open ditches, burying drainage structures, and covering them with soil. Gaining an understanding of how these methods were actually employed to drain wetlands can be helpful in identifying actions needed for their restoration.

The most important step to draining a wetland involves first identifying then blocking its main supply of water. Early landowners recognized that runoff from mountains and hillsides was responsible for maintaining many wetlands, and so they dug ditches at the base of hills to divert this flow.[1] The role of springs in the maintenance of bogs, marshes, and swamps was clearly recognized by authors promoting land drainage, along with the need to control their flow by ditching.[2] Often a single ditch located at the base of a hill or down through the middle of a wetland was all that was needed to dry a site for more useful purposes.

Charles Elliott described two classes of ditches required for draining land: "artificially constructed through swamps, level table lands without adequate natural drainage outlets, river bottom lands or salt marsh lands near the coast; and ditches which are made by enlargening [sic], straightening, or otherwise improving natural streams and watercourse in such a manner as to reclaim and sufficiently protect adjoining land."[3] Clearly, the presence of open ditches shows that wetlands were once present in a particular area.

In a book written to help farmers drain wet land and reclaim eroded areas for crop produc-

tion, Quincy Ayres and Daniel Scoates explain the main object of drainage to be that of the "prevention wherever possible of surplus water from getting on or into the soil as well as the removal from the surface and the interior of the soil of the surplus water that cannot be intercepted." They recommend that open ditches be used to remove surface water and buried clay tiles to remove subsurface water in order to create farm fields.[4]

Individuals have used many techniques for digging ditches to facilitate wetland drainage. Ditches were dug by hand, with animal-pulled scrapers and drags, with engine-driven road graders and dredges, and even by dynamite. Slip or scoop scrapers pulled by horses, mules, and oxen were commonly used to dig small ditches and to move creeks on farms. These scoops were inexpensive and readily available to the landowner.[5]

I asked George McClure to describe the steps his drainage business would take to improve an area for farming, as his success in draining thousands of acres is well known by farmers in northern Kentucky and southern Ohio. He said: "First remove the surface water. Sometimes open ditches are all you'll need. I try to keep the customer's interest in mind. If you can get the water off with ditches, it will save them money. By only doing what is really needed, they'll speak well of you to their neighbor, and you'll get called back for more business."[6] In 1903, Henry French took the same steps to drain lands by first digging open ditches to dry areas, which would also facilitate the possible installation of below-ground drainage structures at a later date.[7]

One of David Murphy's earliest memories

Wooden V-drag with metal cutting edges being used to dig a drainage ditch in Minnesota in 1924. (USDA photograph from Quincy A. Ayres and Daniel Scoates, *Land Drainage and Reclamation* [New York: McGraw-Hill, 1928], 132)

growing up on a small farm in eastern Kentucky in 1962 was watching his neighbor Frank Lewis drain wetlands with a couple of mules and a log. David told me:

Their names were Bob and Barney, I can still see them walking through that deep mud pulling a log. It was a big log—about 14 inches in diameter and 8 feet long and it dug the smoothest trench. He had to make a number of passes to cut out the ditch, but each time that big log would dig deeper in the ground. When he got to the end of the ditch he would move the hook and chain over to the other end of the log and pull it backwards in the same ditch, instead of turning around.

David described the areas being drained:

They were sloughs along the lower edge of fields. Water stood on them, and there were willow trees. One was about a quarter acre, the other a half acre. Frank cut the ditches through the riverbank that had the water backed up. The sloughs grew grass after the water was gone, and he grazed cattle on them. The cattle ate the alders and kept them from growing back.

David also remembers helping his father drain a small wetland on their farm with a mule scoop

and a tractor: "First, we cut the willows out by hand. The slough was narrow, so we could use a long cable to pull the mule scoop behind the tractor. You couldn't go near the slough with the tractor or you'd get stuck. Later, my Dad had Frank come over and clean out the ditch with his mules and the log. Those mules could go places you wouldn't think of putting a tractor."[8]

The Forest Service acquired Frank's farm along the North Fork of the Licking River and Murphy's farm along Upper Lick Fork Creek in the 1970s as part of the Daniel Boone National Forest. The locations that had been drained have since grown up to large sycamore and maple trees.

The Soil Conservation Service (SCS) provided extensive technical and financial assistance to landowners who constructed diversion ditches in the 1950s. However, many ditches were dug out of necessity by private landowners well before the SCS became involved. Over time, these excavated ditches will blend into the landscape, and, if grown up to trees, can appear quite natural.

Diversion ditches were also constructed through nearly level lands to remove surface water and groundwater from wetlands. These varied greatly in width, depth, and length. Excavating a diversion ditch involved moving a considerable amount of soil. The soil removed to create the ditch was not simply piled along the edge of the ditch, as it could form a dam that would flood lands on either side. Therefore, the large amounts of soil removed during construction were generally placed in low areas, or wetlands, near the ditch. This cumulative effect on wetlands was recognized in the Midwest by Robert Burwell and Lawson Sugden: "When you dig a ditch to drain a pothole, you have to put the fill material someplace. Often this spoil is dumped into another pothole. Thus when one pothole is destroyed by ditching, another may be destroyed by filling."[9]

Typically, higher ground is found along riverbanks and streams; this ground is formed by the deposition of heavier sediments under flood conditions, and the rise in elevation forms a natural levee that parallels stream and river channels. Wetlands are commonly formed when the levee prevents runoff from reaching the natural channel. Recognizing that streambanks cause water to pond over fields, farmers cut through them with ditches to give water a direct path to the stream. When they were first constructed, these ditches were typically kept free of woody vegetation, but over time many have grown up to trees and shrubs and now appear natural. The ditches continue to function even with large trees growing along their banks and are sometimes actually dammed by beaver if they have a perennial flow.

Local units of government were formed to design, finance, and construct long, open ditches used by multiple landowners to drain lands. These deep channels were needed by landowners to drain wetlands on more level landscapes. An example of how extensive construction of these community ditches was can be found in *The Drainage Report of Ohio:*

Under the general ditch laws the commissioners of Wood County have granted and laid out 130 drains, averaging 10 miles in length; and the several townships have granted and laid out an equal number, averaging 3 miles in length. During the summer and fall of 1867 nearly 400 miles of county ditches were made, at an average cost of $1,000 per mile, and about 50 miles of free turnpike roads. One of these county ditches is 30 miles long, from 8 to 20 feet at the bottom, and from 3 to 6 feet deep, deepening and widening as it approaches the outlet. The county spent, in 1867, about $500,000 for drainage purposes, and it is considered a good investment.[10]

These residents of Wood County were serious about solving their drainage dilemma, as their disbursement represents close to $6 million in current funds, a major expenditure for any public works project.

In 1891, W. I. Chamberlain explained that some areas of land in Ohio were so level that the only way outlets for tile drains could be obtained was for farmers to cooperate by allowing long open ditches to be dug across their land, paid for by township or county funds: "After such outlet-ditches are dug it would seem to be the height of

(top) Diversion channel on the farm of Darcy Abner in Powell County, Kentucky, built with federal funds around 1955. Darcy is holding the survey rod and using the tape measure to help determine slope along the sides of the constructed ditch. (Photograph taken shortly after construction by an employee of the USDA Natural Resources Conservation Service)

(bottom) Diversion ditch in Bath County, Kentucky, constructed along the base of a hill with help from the Soil Conservation Service during the 1950s. The men are standing in the bottom of the ditch. (Photograph taken shortly after construction by an employee of the USDA Natural Resources Conservation Service)

(top) Outlet ditch on the Raymond Roberts farm in Bath County, Kentucky, dug with a dragline around 1952. The ditch is 7.5 feet deep where it cuts through the natural levee on the banks of Slate Creek. (Photograph taken during construction by an employee of the USDA Natural Resources Conservation Service who assisted with the project)

(bottom) Drainage ditch excavated through a bank on the Licking River in Rowan County, Kentucky. Long grasses and weeds grow in the center of the ditch, and tall trees surround its outlet near the river where it has been too wet to plow.

folly for the individual farmer to fail to get the full benefit of the big ditch he has helped to pay for, get his pay, I say by systematically 'tiling out' at necessary intervals all the land he intends to till."[11]

The Powell County Soil and Water Conservation District in Kentucky acquired two army surplus draglines for creating open drainage ditches to improve farmland from 1950 to 1970. These draglines were efficient pieces of equipment for their time, enabling large quantities of soil to be moved in wet areas where dozers could not operate. The operation of the draglines in Powell County was funded by the federal government in a program designed to improve the quality of life in Appalachia. In 1969, a number of these ditches were constructed by dragline on the James H. Hall farm along Hatton Creek in Powell County. Besides digging ditches to drain wetlands on his farm, the dragline was also used to deepen Hatton Creek so that it would carry more water, serve as a deep outlet for tile drainage, and be less likely to flood fields being farmed.[12]

There are problems with using open ditches to create farmland, as they can encumber planting and harvesting operations. Since most are too deep or soft to cross with equipment, they often end up splitting a field into smaller fields, making it necessary to turn around and back up with a tractor, slowing progress. Soils can be saturated and soft near ditches, making it is easy to get stuck when working alongside them. Ditches also occupy space that could be better used for growing crops. Beaver can block open ditches, often resulting in the need for constant maintenance. Open ditches are also subject to erosion, especially near their outlet. Erosion in ditches washes away valuable topsoil and can further reduce the size of a field.[13]

Problems with open ditches were recognized by the editor of the *Rural New Yorker*, who wrote on May 5, 1855, about draining low land owned by Mr. Samuel Butterfield, at West Cambridge: "Open ditches had been made, but they would fill up by the falling in of the banks; they occasioned much waste of ground, and were an obstruction in working on the lots with teams." He went on to describe how the area was later drained with buried clay tiles that were laid on boards.[14]

In 1861, John Klippart recognized the convenience of using open ditches to drain swamps and water-filled depressions, but he also warned that treading cattle and frost action would gradually fill them. He preferred using buried drainage structures, as they resulted in more thorough drainage with no loss of tillable land.[15]

Beginning in the 1800s, explosives were regularly employed to drain wetlands. Sylvanus Burtis of Ontario County, New York, used blasting to remove rock in order to lay a buried clay tile line in 1857.[16] Using dynamite to create ditches for wetland drainage was described by Charles Elliott in his book *Engineering for Land Drainage,* first published in 1902.[17] Apparently, the practice received considerable application.

In the 1930s, Ricky Wells's grandfather detonated hundreds of sticks of dynamite to create open ditches in parallel straight lines on his land. Buried clay tiles would not perform well in the wetland soils of his Bath County, Kentucky, farm. Known as *bedded soils,* a hardpan occurs from 12 to 18 inches below the surface of the ground. Formed by the precipitation of aluminum and silicate, the hardpan creates a cement-like layer that water cannot penetrate. The hardpan would be expected to rapidly re-form over lines of clay tiles if buried in his fields and would prevent water from reaching the tiles. Dynamite was readily available and an inexpensive alternative to hand digging drainage ditches.[18] Don Hurst of the USDA Natural Resources Conservation Service (NRCS) says that the same technique was also used to drain lands in Powell County, Kentucky.[19]

After World War I, the U.S. Department of Agriculture (USDA) distributed large quantities of surplus explosives to farmers to use in blasting ditches, rocks and stumps and loosening compacted soils. Ayres and Scoates describe in great detail how dynamite containing from 50 to 60 percent nitroglycerin was used by landowners to successfully create both small and large ditches

Deep constructed drainage ditch crossed by Highway 68 in Trigg County, Kentucky.

in swamps holding saturated soils and standing water. They claimed that explosives would work in situations where no other method could do the job and gave understandable instructions on how dynamite could be used to provide a rapid and cost-effective way to maintain ditches dug by dredges.[20]

Richard Bond Jr. was taught how to use explosives by the U.S. Army in the Korean War. After returning home to Carter County in eastern Kentucky, he used this knowledge to create drainage ditches in the early 1950s. "You could cut a 1,000-foot long ditch in seconds," he said. I asked how he would go about constructing a ditch with dynamite. He first used a steel bar to punch holes in the ground to the same depth as he wanted the ditch to be. Most of the time, he would place one stick of dynamite in the hole, but for ditches up to 3 feet deep he used two sticks. The holes were made in a line 1 foot apart. The sticks did not have to be tied together, as the blast from the first would set off a chain reaction causing all the other sticks to explode. Bond agreed with Ayres and Scoates that this was a common, inexpensive way to drain lands, and he added that the new ditch required little, if any, grading.[21]

Piled soil

(top) Earl Stokley operating an Army surplus dragline owned by the Powell County Soil and Water Conservation District to dig a diversion ditch at the base of a hill on the James H. Hall farm in Powell County, Kentucky, in 1969. Note the soil piled in the background along the opposite edge of the field from channeling Hatton Creek at the same time. (Photograph by Roger B. Wiedeburg, USDA Natural Resources Conservation Service)

(bottom) The same diversion ditch on the James H. Hall farm 34 years later. Topographic maps now show the ditch as a perennial stream.

Waterways

Many of the early open drainage ditches dug by farmers were soon eroded by flowing water. Their steep slopes and sides, combined with farming along the edges, resulted in ugly scars across the landscape. The value of creating gently sloped, green, untilled waterway ditches that a plow or cart could pass over was described by John Johnstone in 1808.[22]

Although untilled waterway ditches are generally not thought of as being used to drain wetlands on slopes, Henry French provided a useful illustration on how wetlands were drained on steeply sloped lands:

Again: where land to be drained is part of a large sloping tract, and when water runs down, at certain seasons, in large quantities upon the surface, an open catch-water ditch may be absolutely necessary. This condition of circumstances is very common in mountainous districts, where the rain which falls on the hills flows down, either on the visible surface or on the rock formation under the soil, and breaks out at the foot, causing swamps, often high up on the hillsides.[23]

The SCS began funding programs in the1950s that were aimed at stabilizing eroded ditches by changing them to grass waterways. Grassy waterways were constructed to have a gentle downhill slope, with gradual slopes also being placed on the sides of the ditch to prevent erosion. Waterways were designed so they could be crossed with equipment, an important advantage to the farmer when accessing farm fields. In order to receive cost-share dollars for this improvement, farmers were required to maintain grass in the waterways and not plant them to annual crops.

Private drainage contractor George McClure established miles of waterways for the SCS from 1972 to the late 1990s, mainly in the Ohio counties of Gallia, Jackson, Lawrence, and Scioto. He said that he was required to place a gradual 30:1 slope on the edges of the ditches to prevent erosion and to ensure that they could be crossed by farming equipment. He often buried 4-inch or 6-

Aerial view of parallel drainage ditches created by blasting with dynamite on the Ricky Wells farm in Bath County, Kentucky.

District Conservationist Randy Smallwood, of the USDA Natural Resources Conservation Service, in 2003, next to a drainage ditch created with dynamite during the 1930s on the Ricky Wells farm.

Ditches established with dynamite, now marked by rows of tall green weeds in Ricky Wells's tobacco field.

Constructed grassy waterway near Salt Lick Creek in Bath County, Kentucky.

Grassy waterway constructed in Bath County, Kentucky, blends in well with surroundings.

Waterway constructed with assistance from the Soil Conservation Service, on the Cleveland Brown farm in Menifee County, Kentucky, on May 20, 1957, one year after completion. (Photograph taken by an employee of the USDA Natural Resources Conservation Service who assisted with the project)

inch-diameter plastic drain lines in these waterways to help carry away surface and subsurface runoff. These waterways can be difficult features to detect on the landscape after a few years: their shape is so subtle that they blend into the surroundings as they passively drain wetlands decades after construction. Waterways were most certainly used to drain wetlands in Ohio. McClure is familiar with a number of locations where waterways dried emergent, ephemeral, scrub-shrub and wet meadow wetlands. These wetlands had been located along gentle drainages, at the base of hills and near creeks.

McClure was also required by the SCS to bury drain lines in waterways he constructed on private lands in Ohio for government-funded projects. He buried the drain line a little distance up from the bottom of the ditch at a depth where it ran 1 to 2 feet below the base of the waterway. After government cost-share ended for the practice, it was less common for him to bury drain lines in the waterways, as the expense was paid for entirely by the landowner.[24]

Richard Bond helped dig miles of diversion ditches and waterways for the SCS over a fifty-year career as a drainage contractor in eastern Kentucky. He said that they regularly buried clay tiles, and later plastic drain lines, in the waterways. Surprisingly, he didn't bury drain lines in the very bottom of a ditch but up a little, on one side. Since the bottom of the ditch was often soft and muddy, it was better to dig along the upper side of the ditch. The bottom of the ditch could also wash and erode, exposing the drain line. He buried the tile line 30 inches or more deep, so it was placed well below the bottom of the ditch and drew water the same as if it were located in the bottom of the ditch. He generally stationed the drain line along the downhill side of the waterway.

I asked Bond if there was any way to predict whether or not a drain line was buried in a ditch. He said to suspect the presence of a drain line if the bottom of the ditch was about 12 feet across, because this width was needed for working equipment to install the line. In comparison, a narrow, deep, V-shaped ditch was less likely to contain a buried drain line because it would have been difficult to use machinery in such a place.[25]

Roadside Ditches

One of the most common features on our landscape today is the roadside, which has been used to drain wetlands for hundreds of years. Many roads were originally built through wetlands, and their construction necessitated wetland drainage. According to Allan Studholme and Thomas Sterling, "Large but unrecorded acreages of wetlands have been lost because of road construction. Roadside ditches act as drainage channels, draining one slough into another, or into a creek or river system, where much of the water is eventually lost to waterfowl. Roadside ditches further the drainage efforts of farmers since they serve as channels into which nearby ponds can be emptied."[26]

Roadside ditches gave landowners the deep outlets they needed to install buried systems for draining wetlands, and since roadside ditches were located in public rights-of-way, they provided a legal path for moving unwanted water across a neighbor's property. There are many locations where outlets may be found for buried clay and plastic drainage systems in roadside ditches.

Wetlands are seldom seen along roads because of the long-lasting effects of roadside drainage ditches. Transportation agencies devote considerable effort to the maintenance of roadside ditches by removing soil, debris, woody vegetation, and even beaver dams. These efforts are considered necessary, as water left standing near a road may saturate the roadbed, resulting in significant damage to the base of the road, and failure of the gravel or blacktop surface. It has been found to be easier and less expensive over time to maintain the roadside ditches than to repair or rebuild the roads.

CHAPTER FOUR

Plowing to Drain

Perhaps the least expensive means for growing some crops in a wetland involves plowing soils in such a way as to create areas of higher ground. Soils are directionally plowed in the driest time of year to raise portions of a field high enough so that roots of planted corn, tobacco, wheat, oats, and grass will not drown. Farmers may call the method of directional plowing in Kentucky and Ohio "bedding fields," "making W ditches," or "lands." The practice is used to farm wetlands on clay and silt loam soils with either standing water or a high water table, or both.

On the ground, "lands" look like a series of parallel ditches that have been dug at about 60-foot intervals over a field, with the ditches moving water ever so slowly downhill. Water can stand in the shallow ditches during the winter and spring and, after a heavy rain, most any time of year. In summer, the ditches may appear as rows of bulrushes. Hydric soils are generally present in the ditches and can be found 1 foot or more below the surface of the higher ridges between the ditches.

John Newman and his father before him have labored since 1899 to farm the wet fields along Fox Creek in Fleming County, Kentucky. I met John on an unusually warm January day in front of his now-closed general store along Highway 32. He greeted me by saying, "It took my father 40 years to go broke running this store and me only 20 years to do the same—I guess that's progress." That afternoon I gained an appreciation of how hard he and his family have worked to raise crops on some very wet land.[1]

John showed me a number of areas his father had modified by plowing so they could grow corn and tobacco. He called the farming practice "putting a field in lands." The first thing I noticed was the series of shallow ditches with water standing in them. He drew my attention away from these to point out the slight rise of ground he planted between the parallel ditches. The long narrow ridges were about 30 feet wide and 1 foot higher than the bordering ditches. "You can plant about fourteen rows of corn on a land, but the rows near the bottom usually don't do very well. Some of this ground is so flat you can't drain it with tile. There's no place for an outlet, and you can't run the water over your neighbor's land. Lands are the only way you can farm," he said.

John explained that the best scenario is when the water that runs along a land can be emptied into a deeper ditch or creek, but that this wasn't always possible. Although the practice still works without an outlet, it results in even less area to farm. He said that the speed of a modern tractor allows a farmer to make the ridges of soil twice as wide as he would have been able to with a team of horses years ago. The tractor pulls the plow fast enough that it throws dirt higher than the energy horses could muster.

John looked over the fenced field of lands and indicated that if the field were ever to be used for pasture, his father would insist it be leveled first. Asked what was wrong with leaving it in the lands formation, he said, "Well, it shows you have a wet farm, and that is not a good thing. When you go to sell it, people will know it's wet and it won't bring nearly the price." He explained how he has leveled fields placed in lands by plowing a pattern opposite the direction of the lands, and then discing and pulling a drag at a diagonal across the field.

To make the lands concept more understandable, John sketched the plowing pattern, along with a profile of how elevations would vary over the field. This clearly showed how the practice would only make a portion of the land suitable for crops, but as John had said, that was better than nothing.

Manly Miles described how, by the year 1616, plowing in lands was used to farm wet ground in England.[2] In his 1808 book on draining land, John Johnstone also showed how the practice was being used in England to farm wet soils.[3] Henry French called the practice of farming in lands the "old ridge and furrow system," saying that its acceptance in England was almost universal before tiles were used and noting that it was still sometimes practiced in the United States.[4]

Drainage author W. I. Chamberlain described how he created lands on his Ohio farm: "Then I tried (as already stated) plowing in narrow lands with deep dead furrows for surface-drainage; but this drained off only the water *on* the surface and near the surface, and frequent crop failures, partial or entire, followed. . . . This convinced me that our clayey soils not tile-drained are not fitted for extensive plowing and a successful rotation of crops."[5] John Klippart explained that burying drain lines was superior to the practice of lands, as drainage was improved and the area available for growing crops was considerably increased.[6]

The practice of creating lands in areas where water stood on the surface was called "dead furrow drainage" or "putting small V-shaped surface ditches at frequent intervals" by Quincy Ayes and Daniel Scoates in 1928. They recommended using the technique only until funds became available for installing a system of buried clay tiles.[7] John Haswell stated that the presence of lands provided an indication of the previous wetness in an old field and that this practice was used by farmers who did not appreciate the need to remove groundwater.[8]

George McClure could readily point out areas farmed in lands, or what he called "W ditches," as

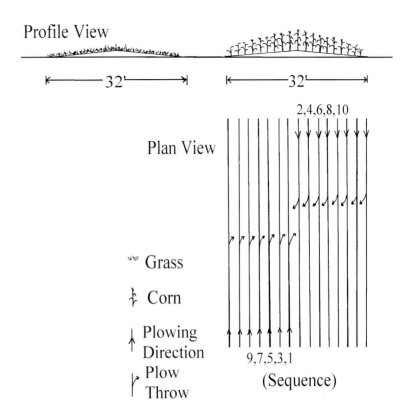

How "lands" were made to farm wetlands. (Drawing by Dee Biebighauser)

he was driving at 55 mph: "Just look for the lines of bulrushes and standing water," he said. He rarely buries drain lines in these areas and found that if drain lines are present, they are generally plugged. George explained that where W ditches are used, land can be difficult to drain for a number of reasons. The most common is that the owner cannot find an outlet deep enough for buried drain lines to function. In addition, sometimes a suitable outlet is located on the property of a neighbor who would not grant permission to cross his land with a ditch or even a buried drain line.[9]

The technique of bedding, or lands, is also used in forestry to grow trees in wet areas. Generally a tractor and plow are used to create narrow ridges of soil on which to plant the seedlings. The practice was used in the 1970s to establish walnut and white pine plantations in the Daniel Boone National Forest in agricultural fields that had saturated soils. These fields recently acquired by the Forest Service had probably been wetlands in the 1800s. The slightly raised parallel ridges and

John Newman on a rise in the ground, known as a land, that he created by directionally plowing a field of saturated silt loam soil near Fox Creek in Fleming County, Kentucky.

corresponding ditches with linear pools of standing water in them still remain visible beneath the planted trees 30 years later.

It appears that some wetlands were drained by the deep cultivation of compacted soils. Long "knives" or "subsoilers" pulled behind a tractor were used to break up a hardpan that held water,

A dark line of rushes growing in one of many shallow ditches created by directionally plowing a field to form ridges for planting near Symmes Creek in Gallia County, within the Wayne National Forest in Ohio.

providing an avenue for runoff to soak into the soil.[10] This deep-plowing activity was effective in some areas in eliminating compacted layers of silt and clay that formed wetlands.

Mole Plow

Designed to cut deep, covered channels in the soil, the mole plow provided an inexpensive way to remove both surface and subsurface waters from the land.[11] In 1858, James M. Trimble used a mole plow to drain a large wetland on his farm near Rattlesnake Creek in Fayette County, Ohio. Trimble wanted to ensure the channels were dug on grade, so he purchased an engineering level before he started work. The level helped him determine how deep the outlets needed to be. First, he constructed 685 rods (2.1 miles) of open ditches 80 rods (1,320 feet) apart, 4 to 6 feet deep and 6 to 8 feet wide, each designed to remove surface water and to serve as deep outlets for the mole plow channels. Next, he staked off lines between the ditches that went through the wettest areas, not worrying about making straight lines. He then used two yoke of cattle, two men, and the mole plow for 16 days to make 1,500 rods (8,250 feet) of under-drain 40 inches deep. The surface of the soil on the areas drained was black clay loam, and the subsurface where the mole plow cut its channel was yellow clay. He ended up draining 230 acres of fairly level prairie land that had from 1 to 6 inches of water standing on it. No one was sure how long the channels would stay open, but the account says that they continued to run water a year after being dug.[12]

A. B. Dickinson used the mole plow to drain from 10 to 20 acres a day where "the ground is more or less springy and saturated with water" with buried channels averaging 3 feet deep and 33 feet apart in Steuben County, New York. The mole plow would not work in rocky soils and was most effective in clay soils, which were less likely to collapse. Dickinson believed that these channels would remain open an average of 10 years.[13] In my opinion, it is possible that some of these earthen drains could flow much longer, as the movement of water beneath the surface often works to keep channels open. Trimble estimated that the mole plow had been used to make over 200 miles of drain in Fayette and Clinton Counties from 1858 to 1860, at the low cost of 5¢ per rod when the work was done by the landowner.[14]

A main concern expressed over using the mole plow to drain lands was its lack of permanence. Tests in heavy clay soils in Michigan found that many filled with soft mud after only one winter, while those created in peat lands within the Florida Everglades were still open after 5 years of use.[15]

Contractor John Utterback of Fleming County currently uses a mole plow to drain wet meadow wetlands on sloped fields in eastern Kentucky. The mole plow he built creates a 4-inch diameter hole, 2 feet deep in the ground, and is easily pulled by a small, 40-horsepower tractor. "It's a fast, quick way to drain wet places for farming; . . . the holes appear to stay open for 3 to 4 years," he says.[16]

White pine plantation established on a drained wetland in Menifee County within the Daniel Boone National Forest. Narrow raised areas between parallel ditches show how the practice of bedding was used to create ground dry enough for seedlings to grow.

The mole plow, developed in 1797 by Adam Scott—an inexpensive way to drain large wetlands on clay soils. (From John H. Klippart, *The Principles and Practice of Land Drainage* [Cincinnati: Robert Clarke, 1861], 232)

Sticks
and Stones

CHAPTER FIVE

Long continuous boxes were often constructed from boards and then buried to drain wetlands in areas where wood was plentiful and clay tiles were expensive or not available. Several authors described how larch boxes or tubes were used to drain swampy or mossy soils in Scotland. The tubes had holes drilled in them and were claimed to be used with economy and good results.[1] Wooden boxes made from green-cut hemlock boards were used to drain lands near Rome, New York, in 1856. The boxes were buried 4 feet deep in wet soils and were found to be in excellent condition 15 years later.[2]

Wood Boxes

Buried wooden boxes were used to drain wetlands and redirect waters from small streams underground in a number of locations in the mountains of eastern Kentucky. John Newman was helping his father install a plastic drain line in their 5-acre bottomland field near Fox Creek in Fleming County, Kentucky, in the late 1960s, when the backhoe operator uncovered a wooden drainage structure. The box was three-sided, with an open bottom. It was made of boards that had been hand-sawn and hewn from chestnut. Its sides measured 2 inches thick by 6 inches wide, with a rounded top covering the sides. They found it buried at the same depth they were digging to place the new drain line, which was about 30 inches. "The box was open and carrying water," said John. "You could reach your arm back in it and feel that it was clean inside." John's fa-

ther, Ronnie Mitchell Newman, who was born on the farm in 1899, said that the farm's previous owners, the Campbells, must have installed the wooden ditch before he bought the farm.[3]

In 1988, after his father's death, John found the same wooden structure again when installing a second drain line across the lower portion of the same field. That time, the contractor exposed the wooden ditch with a trencher he was using to bury the 4-inch plastic drain line. The ditch continued to carry water and the boards were still in good condition. They found that the wooden box emptied into the same ditch that led into Fox Creek near where they placed the outlet for the plastic drain line. The outlet for both drainage structures was a short 3-foot-deep ditch that emptied into Fox Creek, which John speculated had probably formed when the end of the wooden ditch collapsed. He estimated the length of the buried wooden box to be 1,500 feet based on the two remnants found.

John is convinced that the wooden ditch has helped his family grow tobacco and corn in the field for years and that it continues to function today. He would not doubt that more wooden ditches were placed in the area since clay drainage tiles were not available until the Lee Clay Tile Plant opened for business in Morehead. The plant began operations in Clearfield, a community near Morehead, in 1926 and capitalized on the high-quality clay found on mountain ridges in the area.[4] John said that Fox Valley is known for being a wet area, and farming its bottomlands has taken considerable drainage action over the years. As we looked over the large field where the wooden ditch lay, asked to describe how it would appear without the drainage structures, he said, "I would not be able to farm without that drainage; why, water would stand on it."

Retired district conservationist for the NRCS John Meredith also remembers finding buried drainage ditches constructed of wood boards in Athens County, Ohio. He claims they were not at all common and guessed they were probably installed before clay tiles became available.[5]

Ed Stevens in Carter County, Kentucky, proudly showed me a field where he had buried 3,000 feet of plastic drain line so the acreage would be dry enough to raise corn the next summer. As we looked over the lines of backhoe-dug trenches, he recalled as a boy hearing old-timers talking about burying three-sided boxes made of chestnut boards to drain lands. He wasn't aware of the location of any of the fields drained by this method but said that, because chestnut was so resistant to decay, he would not doubt the underground ditches lasted a long time.[6]

Brush Ditches

A number of authors describe burying brush and small-diameter trees in ditches to drain swamps, bogs, and other wetlands.[7] Henry French tells that J. F. Anderson of Windham, Maine, used small-diameter trees to drain lands and recommended the practice "in regions where wood is cheap and tiles are dear." He speculated that pole drains would be nearly imperishable in soils that were constantly wet.[8]

Richard Bond remembered encountering four or more of these brush drains over the span of his career in eastern Kentucky and that all were still working when he dug into them. He began operating heavy equipment to make a living after serving in the army during the Korean War. His first job after returning home to Carter County in 1953 involved constructing ponds for farmers with a dozer in cost-share projects funded by the Soil Conservation Service. A few years later, he used cash to buy a backhoe and began helping

Approximate location of a wooden box drainage structure, buried in the 1800s, that continues to function on John Newman's farm in Fleming County, Kentucky, indicated by the red line. About one-half of the ditch's 1,500-foot length is shown.

farmers drain their lands. They hired Richard and his backhoe to drain wetlands with open ditches and clay tiles for a number of years.[9]

In the early 1970s, Richard again used cash to purchase a tiling machine that he eventually used to install hundreds of miles of buried plastic drain lines in a fifteen-county area encompassing eastern Kentucky and southern Ohio. While installing these clay, and later plastic, drain lines, it was common for him to encounter early buried drainage structures. These drains were made from rock, wooden poles, and wooden slabs. When questioned about the effectiveness of these primitive structures, he replied, "Those guys knew what they were doing; they were buried at the same depth as I was at, headed in the same direction, and still carrying water. I'd have to tie into them with my line or there'd be a wet spot in the field."

Richard and I traveled out from his home to look at a couple of fields where he remembered finding buried rock and wooden drains. When asked his theory as to when these things could have been buried, he paused, thought for a while, and said, "I can't tell you when they started using them, but I can tell you when I think they stopped.

A large field, once a wetland, in Powell County, Kentucky, originally drained by James H. Hall in the 1930s with hand-buried hollow logs and triangle-shaped board structures. The puddle that remains is one of only two sites in the county where spadefoot toads breed.

It was when Lee Clay was built. They were the first to make clay tile in this area."

Richard and I drove into Boyd County and turned up the Bear Creek Road, keeping our eyes open for a small, two-story store that he remembered being opposite the field he was looking for. He spotted the store with ease, even though it had since been converted to a home. We parked in the driveway of the person who once owned the field and found that no one was home. He explained that the owner hired him to drain a wet place in the field sometime in the early 1970s.

Richard described how he was using his backhoe to start a ditch near the lower edge of the field along Bear Creek and dug perpendicular to the creek up into the field. He then turned to follow a shallow dip through the middle of the field that paralleled the creek. As he was digging up through the middle of the field, he unearthed a long pile of wooden poles, from 2 to 4 inches in diameter, that had been buried in pyramid-shaped piles of six, in a line at the same depth he was digging. He had a lot of trouble digging a trench through the poles with the backhoe because they were staggered in the ditch, and some were 16 feet long. The poles

were sound; bark was still attached, and water was flowing down the ditch between spaces toward the creek. He made a T in the plastic line he was installing, to connect in with the pole ditch, and covered the junction with creek rock. Richard continued digging the ditch and installing the 4-inch-diameter plastic drain line until he reached the wet place, which had been located uphill from where the pole ditch was headed.

The area with the pole ditch is a dry, 3-acre hay-field located below a blacktop road and up from the creek. No wet spots or clumps of bulrushes were visible in the field, and this had been one of the wettest years on record. The field was not level and appeared to drop about 4 feet over its length. Water was being kept off the field in a number of ways. The roadside ditch diverted runoff before it could reach the field. The intermittent stream coming off the hill had been straightened and turned into a ditch that now bordered the southern edge of the field. And, of course, the pole ditch combined with plastic drain line worked to keep water off the area.

While examining restored wetlands at the Wayne National Forest in Ohio, I was discussing how much work it must have taken to drain the fields in the area for farming with Richard Neal, who was grading the road on which I was parked, when he mentioned that one of the fields he farmed across the creek had been drained by burying wooden poles in the ground. Just north of the little community of Lecta, he owned an 8-acre field planted to tobacco and hay. Richard had been told by his father that in the early 1900s, a series of ditches had been dug in the field that emptied into the creek. Rows of trees from 3 to 4 inches in diameter were laid in the ditches and stacked in triangle-shaped piles of three. The tree bundles were then covered with soil to form underground passages for removing excess water. This process had been used because the farmer could not afford clay tiles. A field that had been too wet to farm has since produced corn, hay, and tobacco for generations. Richard believes that the drainage structures are still working, as there are no wet spots in the field.[10]

I spoke with John Meredith a few days after he retired from a 36-year career with the NRCS as the district conservationist for Jackson and Vinton Counties. He recalled uncovering drainage ditches made with buried wooden poles in Jackson County, Ohio. Calling these "brush ditches," he says they were the most common of the early non-clay drainage structures he had encountered while improving drainage systems for farmers.[11]

For over 33 years, George McClure has installed agricultural drainage systems in southern Ohio and northern Kentucky. He remembers finding a number of brush ditches over the years. The buried trees would be from 4 to 5 inches in diameter and in good condition. Generally, the ditches were still carrying water. He recalls that the brush ditches were difficult to cut through and basically served as a nuisance, as they slowed his progress in burying new plastic drain lines.[12]

It should come as little surprise that buried wood ditches can carry water for many years. John Johnstone in 1808 reported that drains filled with wood were preferable to any other kind of material, as water would continue to pass in the subterranean void even after the wood decayed. He found that covered wooden ditches and buried willows remained in excellent condition for at least a 30-year period.[13]

Wood Slab Ditches

Henry French describes how wooden fence rails, sod, and straw were used to create buried drainage ditches near Washington, D.C. He said that these worked best in clay soils compared to sandy soils, where they were more likely to become filled.[14]

Richard Bond remembers uncovering wooden slabs that had been buried in fields to drain water. The slabs, which appeared to be oak, were from 1 to 2 inches thick by 4 inches wide and were buried in stacks in the bottom of ditches about 2 feet wide. The slabs were in lengths from 8 to 16 feet long, laid staggered in the ditch, placed with alternating bark up and down to leave space for water to travel. I asked Richard why they dug

Richard Bond uncovered a functioning, buried drainage ditch constructed of 4-inch-diameter wood poles that moved waters to Bear Creek, located at the base of the hill in Boyd County, Kentucky.

the ditches so wide, and he said, "You can't dig a ditch any narrower and still get in it. You know, they dug these out by hand and had to have room to work." Richard thinks they started using slab wood in the 1920s when sawmills were common in the area.[15]

Cropped field from a wetland drained in the early 1900s by hand-burying trees in ditches along Sand Fork Creek in Gallia County, Ohio, within the Wayne National Forest.

Rocks and Ditches

The construction of buried rock drains for drying wet spongy ground in Ireland was described by Sir Charles Coote in 1801:

These drains have been found very serviceable, soon making a wet spongy soil dry and sound, and able to bear cattle in winter. A drain is cut about two feet and a half deep, inclining to a slope at both sides downwards, from eighteen inches at top to six inches at bottom; this trench is filled with the largest stones, forced in between the sides of the drain, and covered in with paving stones, then a layer of brush-wood, etc. to support the clay, and about twelve or fourteen inches from the surface, over which is thrown the clay, that was dug out of the drain. This method is effectual, and, where stones are easily had, is very cheap.[16]

Henry Stephens in his 1847 *Book of the Farm* explained how he drained a 2-acre ephemeral wetland that was surrounded by a 25-acre field. The water-filled area had been nicknamed the "Duck-mire" as wild ducks frequented it every season. He began the project by hand-digging a 10-foot deep ditch about 150 yards long from a clay bank above a small river over to the edge of the wetland pool. In the bottom of the deep ditch, he constructed a rock channel measuring 9 inches wide by 12 inches high. He then placed 2 feet of clean stone over the rock box, covered the stone with turf, and filled the trench to the top with soil. The next step involved digging a ditch through the center of the wetland. The bottom of the ditch was kept 30 inches lower than the bottom of the wetland in order to maintain a downhill slope toward the river. He had a great deal of trouble completing the ditch through the wetland, as it traversed quicksand and its base was too soft to support the rock channel. To solve this problem, he cut thick sections of sod and placed them in the bottom of the ditch. He then laid flat stones on top of the sod to form a foundation for the rock box and the loose stones piled on top. Stephens carefully packed sod against each side and over the top of the rocks to prevent sand from flow-ing into the structure. The entire arrangement was then carefully covered with soil. To make sure that the area would thoroughly dry, he built a 3-foot-deep rock and stone channel around the entire wetland, at an elevation a little above the water mark, with the ends of each ditch emptying water into the main ditch, which, in turn, carried water to the river. The following spring, the blue clay soils were dry enough to plow and plant, and "ever after bore fine luxuriant crops."[17]

In 1763, Joseph Elkington developed a system for draining swamps and bogs that created quite a sensation throughout England and Scotland. Described as an illiterate Warwickshire farmer, Elkington discovered how to locate, tap, and remove the water from springs that kept land wet.[18] He would basically dig a trench, from 3 to 7 feet deep, at the base of a hill along the upper edge of the wetland. He found that such trenches were rarely deep enough to reach the source of the water, which was generally a spring. He then used an iron auger to drill holes another 5 to 10 feet lower in the bottom of the ditch.[19] Water typically burst up through the holes, only to be directed downhill in a covered rock ditch that he constructed which led to an outlet, which was often a river, a stream, or an open ditch.

Elkington was said to have great talent in locating underground water and the main source of a spring. Word of how he had successfully drained his and his neighbors' farms spread, and he spent the next 30 years helping others drain thousands of acres of bogs, marshes, and swamps throughout England.

The House of Commons awarded Elkington £1,000 in 1795 for information concerning his mode of draining. In consideration of Elkington's failing health, England's Board of Agriculture assigned John Johnstone the task of recording his techniques in a book that later required three printings.[20] Johnstone wrote after careful study of Elkington's actions in 1796: "According to these principles, this system of draining has been attended with extraordinary consequences in the course of Mr. Elkington's practice. . . . By it, not only the land in the immediate vicinity of the drain, but also springs, wells, and wet ground, at a considerable distance, have been made dry, with which there was no apparent communication."[21]

Elkington's technique for controlling the water from springs was brought to the United States and successfully used to drain swamps, marshes, and bogs.[22] Henry French said that "the Elkington method cut a drain deep into the seat of the evil, and so lowering the water that it may be carried away below the surface, is obviously the true and common-sense remedy."[23]

I. Whiting described how in 1839 he constructed a stone drain that was 660 feet long and 3 to 4 feet deep in New York State. He took great care to make sure the ditch contained a 6-inch drop over 297 feet of its length when draining level ground, and he claimed the ditch was still working 14 years later.[24]

Alvin Wilcox wrote about the farm he owned in West Bloomfield, New York, in a letter dated December 14, 1855:

I have about eighteen acres of land, three-fourths of which was considered good grain land; the remainder was wet land, made so by numerous springs and swales. About sixteen years ago, I commenced draining with stone. Seeing the improvement it made and the extra yield the drained land gave over that which I considered good grain land at that time without draining, I kept extending the stone drains for several years until I commenced manufacturing drain tile and drain pipe.[25]

John Klippart reported, "Considerable draining with stone has been done in Ohio."[26] He describes how stone drains were made at depths from 3 to 5 feet below the surface, and how the use of stone was popular, as materials were obtained for free during "odd half days." Their use also improved the quality of a field by providing a place to do away with stones littering the surface of a field. An example is given where the Honorable John Howell of Clark County, Ohio, had 1,000 rods (3.1 miles) of stone drains on his farm, 28 to 30 inches deep, and filled to a depth of 10 to 12 inches with stone.[27] Henry French rec-

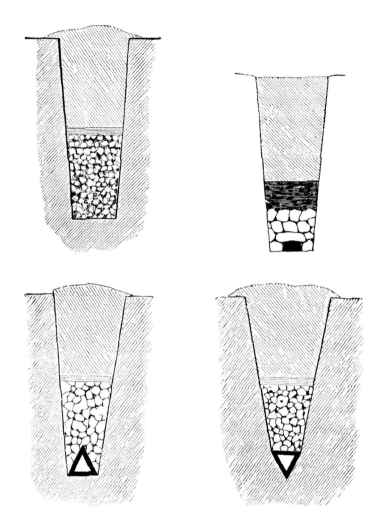

Rock ditch designs by Henry French. The triangles probably represent wood channels, the thin lines show overlying rock sand and gravel, and the thick dark layer may have been sod or topsoil. (From Henry F. French, *Farm Drainage* [New York: Orange Judd Company, 1903], 115)

ommended constructing stone ditches to be at least 21 inches wide from top to bottom, and said that at least two ox-cart loads of stone would be needed for each rod built.[28]

Lost Creek Rock Ditch

Richard Bond guided me up the hollow along a narrow winding road to a place he helped drain in Carter County, Kentucky, on the farm owned by Homer Ratcliff, now deceased. As we looked down over the green valley, he described how Homer had hired him to do many drainage projects over the years, but he remembered one in particular from 1975.[29]

Homer had asked Richard to drain a 3-acre pool of water about 4 inches deep from the field near his barn so he could raise more corn and tobacco. Richard looked the area over and planned to bury a 4-inch-diameter plastic drain line right through the center of the wetland to dry it out. For an outlet, he started digging a ditch at the same level as the bottom of Lost Creek, near the downstream edge of the field. Just as his ditch began to enter the standing water, he began to uncover rock, lots of rock. Now, fields in eastern Kentucky are generally not rocky, consisting mainly of fine-grained silt loam, so hitting rock like this was unusual. The rocks were of various sizes; however, all were small enough to be carried by hand. They had been placed in a covered ditch about 2 feet wide by 1 foot high. Homer had no idea that this rock drain was on his farm before Richard's work revealed it.

Richard uncovered thousands of rocks as he dug in the same course as the rock ditch: "There were rocks all over the place; . . . I couldn't cover them all up." The rock ditch was still carrying wa-

ter, so he added a T to tie into it with the plastic line, covering the area around the junction with rock. His new ditch eventually parted from the rock ditch as he dug into the upper reaches of the wetland. Richard said: "Had we known there was a rock ditch in the field, we might have been able to save some work by finding its outlet near the creek. Maybe it had plugged with some leaves, and we could have cleaned it. This could have saved him some money." Homer did not receive any funding from the government for the project.

Richard said that he ran into a number of these rock ditches over the years while installing drain lines. Rock was always buried in a channel about 2 feet wide by 1 foot thick. The rocks were never worked by hand, and they looked like they had been picked up from areas around the field. The bottom of the rock-filled ditches was always about the same depth as the drainage ditches he was digging, and they all still carried water. He said that the rock ditches had been put on grade, showing that the early farmers were experienced in draining fields. Had they left a low spot, or swag, in their ditches, the rock drains would have filled years ago. They were "as good at cutting grade as us," said Richard with respect. The rock ditches were generally covered with 2 feet or more of soil: "You could run over them with a tractor or dozer and they'd keep working—nothing would hurt them." Richard acted surprised at my interest in these rock ditches. "After all," he said, "look how they're still using them today to drain roads and septic systems."

Ratcliff's Rock Ditch

Eddie Ratcliff walks over a drained wetland each time he visits the barn to feed his cattle. A buried channel constructed of field rock was used to convert the wetland into a tobacco field over 100 years ago in Carter County, Kentucky. The half-acre wetland was located in a depression fed by both mountain runoff and spring water that was prevented from reaching Lost Creek by its higher bank.[30]

A functioning buried rock ditch was discovered in this field by Richard Bond in 1975 in Carter County, Kentucky, while adding plastic lines to drain a 3-acre ephemeral wetland owned by Homer Ratcliff. The straight line of vegetation in front of the buildings marks an open ditch constructed to carry water from the once-meandering stream that flowed over the field straight into Lost Creek.

"My grandfather told me the Conways dug the rock ditch when they built the barn in 1890," said Eddie. "He bought the farm from the Conways about 10 years later. My father showed the outlet to me; it's always had water coming out of it. The ditch is about 1 foot by 1 foot in size, and made from flat rock collected in these hills. I was told that the ditch is 5-foot deep in places, and this must be true because I've never hit it with my plow." His observations of the depth were important, as Henry French found that rock drains constructed more than 3 or 4 feet deep were comparatively safe from blockage.[31]

Eddie showed me where the rock ditch began in the lowest part of the now dry depression, and where it ended at the creek. Under his direction, I measured the length of the ditch at 90 feet. We jumped down to the creek bed and could see water flowing from a vertical face on the exposed creek bank, about 3 feet down from the top, where he said the outlet was located. No rocks were visible from where the water emerged. Eddie said, "The creeks changed channels here about 20 years ago and covered the outlet, but it keeps working." Eddie had also dug two short open ditches near the lower edge of the small field to help remove surface water after the creek had changed channels.

I asked Eddie why the Conways went to all the trouble of making this rock ditch. He explained, "They had to. Surface water will scald tobacco roots in 24 hours or less, and kill the plants. If you can get the surface water off, low ground like this does real good in a drought—it keeps enough moisture to keep tobacco growing all summer."

When I questioned Eddie about the value of draining such a small area, he responded, "This is a real good place to have a tobacco field. It's close to the barn for spreading manure and you can hang it easy. There aren't many good fields on this farm." I found out that his 312-acre farm had only 20 acres of bottom or hilltop fields that were level enough to raise tobacco. Asked what the field would be like without the rock ditch, he said, "Would be nothing but a swamp if that ditch wasn't working."

Eddie said that his grandfather, his father, and he had all raised good tobacco with large leaves from the drained wetland. He explained how the government's quota for tobacco was once based on acres, not pounds. That made it worthwhile to drain even a small field to increase its yield so more money could be earned from the limited ground. He said that there were no clay tiles in that field, but he did have them in a field below the house and thought that Richard Bond may have been the one who installed them.

He last farmed the field 5 years ago and then only to raise watermelons. Reductions in the tobacco quota over the last several years have been large enough that he does not raise tobacco anymore. Now that he is 65 years old, I asked him what he thought would happen to his farm in another 20 years. He thought for a while and answered, "It will be sold; my children don't have an interest in farming."

Walking over the small field, I found it to be well drained, with no aquatic plants growing in it. The original field had been about 1 acre in size, which included the slope at the base of the mountain, a depression, and a rise along the creek. The field was now growing up to blackberry and elderberry, with fescue, goldenrod, and asters covering the ground.

The field had a slope of about 1 percent overall, with a 0.5-acre depression in the middle between the base of the hill and the bank of Lost Creek. The creek bank was about 2 feet higher than the depression. The dirt road leading to the barn followed the rise along the creek, apparently because it was sandier and had better-drained soils. There were no signs of gravel on the access road, and it looked as though it had once been plowed along with the rest of the field.

I dug two soil test holes in the lowest part of the depression and observed brown-colored soils (10YR 4/4 on the Munsell Soil Chart) and no mottles. Water did not enter the holes, even though it was in the middle of winter and we had been receiving above-average precipitation.

Other actions had been taken to help keep the field dry over the years. Spring water and runoff flowing down the hill were being kept from reaching the field by a diversion ditch that ran along the base of the mountain, which also formed the upper edge of the field. A perennial stream that flowed down the hollow formed another edge of the field. It looked as if this stream had been straightened and shifted to make the field larger.

Digging with my shovel, I attempted to expose the rocks that formed the outlet for the rock ditch. The flow from the creek bank increased with each shovel of soil I removed, washing downstream like a garden hose opened halfway. I eventually dug into the bank about 3 feet, but did not hit rock.

Morgan County Rock Ditch

Scott Manning was plowing a 0.5-acre tobacco field in Morgan County, Kentucky, with his new tractor and plow when he unearthed a stone box that was buried along the upper edge of the field. A channel had been made of flat rocks, each squared by hand, measuring 12 inches tall by 12 inches wide. Scott was surprised to see these stones as nothing like them had previously been found on his farm. The continuous box had been buried at the base of the hill beneath about 18 inches of silt loam soil for a distance of about 120 feet from a spring to a little creek near the edge of the field. Scott continued plowing the field and did not take time to repair the stone channel. The tobacco did poorly in the field that year and drowned in subsequent years. Scott stated that much of the field remained wet continuously, and that a good set of waders was needed to walk near the base of the hill. He's stopped raising crops in the field, and it has now grown up to sedges and bulrushes.[32]

Scott remembered that in 1960, when he was only 8 years old, his grandfather worked on the same field to improve its drainage. His grandfather had uncovered the stone drain when he was burying a line of clay tiles. Scott recalled seeing water running inside the stone drain box at that time. His grandfather connected the clay tile line into the stone drain and also ran the clay tiles

to the small stream along the edge of the field. Scott's grandfather, Estil Manning, purchased the hilly farm in the early 1930s with the stone drain already in place. Scott said, "There are few places flat enough to grow tobacco on the farm so they had to drain these small fields to make a living."

John Meredith would occasionally find sub-surface rock ditches when assisting farmers in the installation of drainage systems in Benton and Jackson Counties in Ohio. "They seemed to use these for spring developments," he said. "I'd also find them around home sites where they were particular about drainage."[33]

George McClure was building a pond in a wet place on a terrace just above a gentle hillside in Scioto County, Ohio, when he uncovered a functioning buried rock ditch originally constructed to drain the field. He was making the core beneath the dam when he hit an abundance of field rock, buried 3 to 4 feet beneath the surface. To make sure that the pond would hold water, he used his dozer to cut through all the rock and then packed the core with clay. "Goose Creek is an old German community," he said. "Years ago they raised corn on these hillsides because the bottomlands were too wet to farm. That was before government programs helped them drain the bottomlands."[34]

Examples of fields being drained by buried rock and wood illustrate how committed early farmers were to creating and improving croplands in the mountains and hill country. They used primitive structures to drain wetlands, and many are still flowing today. The early use of buried wood and rock ditches shows that dry areas of land, whether level or sloped, may have been caused by intensive drainage actions completed hundreds of years ago.

Unfortunately, the wetland builder will rarely find documentation of people installing these rock and wood structures, or any outward signs of their presence on a land tract. When working to restore a wetlands hydrology and to ensure its success, it would be wise to dig deep into the ground around the lower edge of each construction site to look for and block drainage structures that are still operating successfully.

CHAPTER SIX

Miles of Tiles

John Johnston introduced the practice of burying clay tiles to drain wetlands for agriculture in the United States in 1838. The use of clay tiles increased greatly in the mid-1800s and became the standard for land drainage until they were replaced by plastic in the early 1970s. W. I. Chamberlain stated: "Tile drainage has superseded all other kinds of underdrainage, as, for example, that with poles, rails, slabs, brush, cobble stones, or with the mole-plow. It is immensely better than any of these; more durable, more efficient, and really cheaper in the long run."[1]

Buried clay tiles work as covered ditches to remove both surface water and groundwater from wetlands. These drain lines can prevent water from standing in fields and lower the water table. Properly installed clay tiles require little, if any, maintenance and are known to keep functioning for years. In 1893, Manly Miles wrote that land draining should be considered a permanent investment.[2] Henry French, perhaps the most widely read drainage author, stated in 1903, "On the whole, after wide observation, I am prepared to assert anew, that tile-drainage will endure forever if the work be properly done."[3] Chamberlain claimed: "Properly made, burned and laid, there is no reason why the tiles and drains should not last for centuries."[4]

Chamberlain said, "Surface drainage is better than none; but it greatly interferes with all farming operations. If the surface drains are natural, that is, simply made by water action, they will usually be crooked brooks or gulleys, cutting up the field into awkward shapes for cultivation."[5] Therefore, it is easy to see why burying tiles became the drainage method of choice for most farmers. A farmer can plow over drain lines, and they work all year, not freezing because they are deep underground. Since they lower the water table and eliminate saturated soils, a farmer can plow and plant earlier in the spring. The excess water is able to leave the soil, not drowning the roots of planted crops. Mr. Martinelli of Nerac, France, described how important the hole is in the bottom of a flowerpot to allow excess water to drain and not drown the roots. John Klippart stated, "The drainage of tillable land is a small hole at the bottom, just like that flower pot."[6]

William Johnson of Geneva, New York, was given a $10 award by the New York State Agricultural Society in 1855 for describing how he converted his wetland into a field:

I have on my farm about eighteen acres of flat low land being a sort of low basin for the deposit of the water running from a large tract of surrounding lands. The soil is a kind of vegetable mold interspersed with clay, with a clay subsoil. Ten years ago I purchased the farm on which I now reside. At that time this piece was overgrown with small trees, bushes, willows, bog-grass, &c., presenting a most unsightly appearance, and was considered almost a nuisance, in fact it was known and pointed out as *the swamp*. In spring after I came in possession of it I cut down all the trees and bushes, burned them, together with a large quantity of old logs, tree tops, &c., then dug an open ditch two and a half feet deep through the lowest part of it, which carried off a considerable portion of the *surface* water, and was really a great improvement, but was not what the land required (nor what I intended to do as soon as more

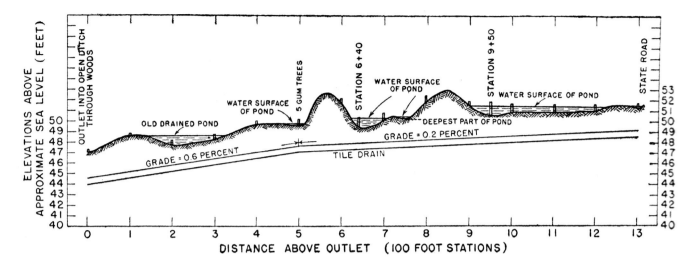

Profile of a 1,300-foot-long, buried 4-inch-diameter clay tile line used to drain five wetlands containing surface water. (From John R. Haswell, "Drainage in Humid Regions," in *Soils and Men: USDA Yearbook of Agriculture* [Washington, D.C.: U.S. Government Printing Office, 1938], 734)

pressing improvements were disposed of), it being a rough uneven piece, full of holes, with a close tenacious subsoil, the water standing in the low places a considerable portion of the year, and of course too wet to be tilled with any success. Last spring I commenced the work of underdraining it in earnest, by cutting a ditch along the east and lowest side of the lot for the *main drain,* thirty inches deep, to be laid with six inch tile. I then commenced on the north and lowest end of the lot with the cross drains, making them about thirty-two feet apart, (varying a little according to the situation of the surface,) nearly at right angles with and entering into the main drain. Now for the result—as the drains progressed the water began to disappear from the surface, and within about one week after the drains were dug the water *entirely* disappeared from the *lowest* places. The effect was striking and remarkable to everyone who witnessed it. That portion through which the drains had been cut being entirely dry, whilst the other portion immediately adjoining, was literally soaked in water, and as fast as the drains progressed the water would as rapidly disappear. The experiment has proved *entirely* satisfactory, and I have already plowed about one third of the lot and intend to plant the whole of it to corn next spring; in fact I expect after it shall have been thoroughly tilled it will be one of the driest lots on the farm, and if the season proves favorable, I have no doubt the corn crop will tell well next year. The actual

amount expended in draining the nine acres described above is $234 . . . about $26 per acre.[7]

Further proving that early farmers used clay tiles to drain wetlands, this observation by Chamberlain is provided: "All over the rolling prairies of Iowa and bordering States, and even of the sandy loams, are 'swales' or 'sloughs,' and 'cat-swamps,' or small wet 'pockets' that need perhaps one or two good four-inch drains put through them to make them arable and most productive. Without such tiling they produce little but swamp grass."[8] When draining "marshlands," Manly Miles recommended that open ditches be dug to remove excess water, and then, after the ditches had lowered the water table and organic soils had settled, that clay tiles be buried 4 or more feet in the ground so that the area could be farmed.[9]

The use of buried clay tiles to drain wetlands was extensive in the 1800s, as described in detail by Mike Weaver in his book *History of Tile Drainage in America Prior to 1900.* Weaver spent most of his career with the SCS designing and installing drainage systems on farms in New York and commonly found evidence of earlier clay tile installation.[10]

George McClure learned how to drain areas by laying clay tiles on his father's farm in Lawrence

County, Ohio, in the 1950s. He started a land drainage business in 1972 that he continues to operate today in Ohio and Kentucky. When asked how often he encounters clay drain tiles when installing plastic drain lines, he replied, "I hit clay tiles in about half of the drainage jobs. About one in eight times when I find them they are the old style." The clay tiles he finds most of the time are round with 4- to 6-inch-diameter openings. The older styles he finds can be arch shaped, or round, with openings from 2 to 8 inches wide. "About half the time they're still working and I'd have to tie into them," he says. George has even found arch-shaped clay tiles without a bottom in Ohio, these being similar to those made by hand in New York before 1850.[11]

Farmers began using buried drainage structures to dry wetlands on sloped lands in the 1800s. In 1871, the Honorable Josiah Shull of Ilion, New York, gave a report at the New York State Agricultural Society that described how he worked to drain a "springy side hill" for farming. Such an area would most likely be called a wet meadow wetland today:

This hill on his farm had been drained with board drains, running from the top to the bottom of the hill, but the ground was still wet. He therefore, cut drains longitudinally and diagonally along the hillside, averaging about three rods apart. These drained every point. He found the soil in the neighborhood of the springs a quick-sand; it was necessary to put in the tile the same day that the ditch was dug.[12]

The account continued to describe how a wet side hill with a 20-degree slope was also drained.

Many early farmers were tenacious in their efforts to drain wetlands, as evidenced by this description of clay tile installation by French:

Messrs Shedd and Edson, of Boston, have superintended some drainage works in Milton, Mass., where, after obtaining permission to drain through the land of an adjacent owner, not interested in the operation, they could obtain but three inches fall in one hundred feet, or a half inch to a rod, for three quarters of a mile, and

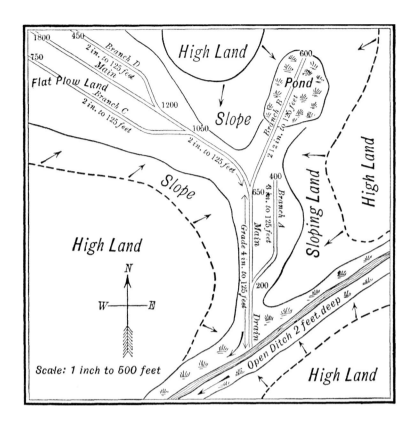

Farm drainage plan from the 1800s showing the location of buried clay tiles (labeled "Main Drain" and "Branch A–D") and open ditches. (From Charles G. Elliott, *Practical Farm Drainage* [New York: Wiley, 1903], 61)

this only by blasting the ledges at the outlet. This fall, however, proves sufficient for perfect drainage, and by their skill, a very unhealthful swamp has been rendered fit for gardens and building lots.[13]

James Flowers's Drainage Experiences

One cloudy November day, my friend and farmer Dewice Copher and Philip Annis, Dewice's business partner, and I traveled to western Kentucky to meet with James "Booster" Flowers, Philip's distant cousin, to talk about wetland drainage. We discovered that Booster's passion for drainage began as a young boy growing up near Kentucky's Green River, where he drained puddles around the barn by using small pieces of cane as imaginary drain tile. He is now 89 years old, and his stories of wetland drainage paint a picture of just how greatly the landscape was changed in But-

Bright green field on a 3 percent slope in Powell County, Kentucky, that was once a spring-fed, wet meadow wetland. The field was made possible by the construction of a diversion ditch and the installation of clay tiles in the 1950s.

Field created on a historic spring-fed hillside wetland with a 3 percent slope in Carter County, Kentucky, drained by Richard Bond Jr. in the 1960s by burying 4-inch-diameter clay tiles.

ler and adjoining counties along the banks of the Green River.[14]

Booster earned 50¢ a day in the 1930s hand-digging ditches for clay drain tile. I asked him to describe how he would go about draining a wet piece of ground. He stated, "First, I had to find a good outlet. This could be the river or a slough. I'd start digging my ditch at the outlet and go 50 feet at a time. I'd use a gopher to finish the bottom of the ditch." He explained that a slough, or gut, was a deep, low place that may or may not hold water. He used a shovel to dig the ditch and a gopher tool, otherwise known as a crum, to cut the groove for the tile to lie in and to remove loose soil from the ditch. "Sometimes we used a 10-inch breaking plow pulled by a horse to help dig the ditch," he said.

The water coming into the ditch told me how much rise I needed for 50 feet. I wanted the ditch to go up a half an inch every 50 feet. The water would be in the tile at the low end, and at the bottom of the tile at the upper end. No one had a level, so the water in the ditch gave me the grade. The tile wall was a half-inch thick, so this provided the rise I needed for 50 feet of ditch. I did the same for every 50 feet of ditch until the ground was tiled.

I asked Booster if he ever poured water in the ditch to make sure it sloped downhill. He gave me a puzzled look and asked, "You really haven't done anything like this?" I shook my head no. He explained that he often stood in water while digging the ditch and the ground was saturated, so water running down the ditch showed him it would drain. Booster buried clay tiles from 1.5 to 4 feet deep, depending on how level the ground was from the outlet.

Now I could tell that Booster wanted to just sit and talk, and we all could see it was getting darker and beginning to rain outside. We had bought his breakfast, and the server was doing a good job of keeping his coffee cup full. He had three people at the table listening to his stories and was in no hurry to go out and look at muddy fields he left years ago.

He made sure I understood how important it was for the foot-long clay tiles to fit snug against each other in the ditch, so that gaps did not remain at the top or bottom between the tiles. "You'd best tile as tight as you can; I'd use a stick to tamp tiles against each other," he said. Soil was then carefully

Types of clay pipes buried for drainage. *Left to right:* arch (7 inches wide by 9 inches tall), round (5.5-inch diameter), round (4.5-inch diameter), octagon (4-inch diameter), horseshoe from John Johnston's farm (1.75 inches wide by 2.25 inches tall), arch (2.5 inches wide by 3 inches tall), round junction (4-inch diameter), and corrugated modern slotted plastic (4-inch diameter).

packed in layers over the top and sides of the tiles to keep them in place. Water entered the drainage line along the underside between the narrow cracks remaining between each clay tile.

Booster had investigated a number of deep caverns left in fields after flooding, only to find clay tiles in the bottoms of these large holes. He would always find a gap between the tiles that had allowed water to enter directly from the surface. Apparently, rising floodwaters would push air back into buried tile lines, and the compressed air would escape to the surface through the gap left between tiles, thereby creating a path for water to gush back down into the tiles and creating a large washout in the process.

The clay tiles Booster used were 4 to 8 inches in diameter, manufactured in Daviess County, and shipped up the Green River. "I would carry them off the barge four at a time by sticking each arm through two tiles, and grabbing the stack at the bottom," he said.

In 1961 Booster began working for the SCS as an aide, and he spent his career designing and inspecting drainage systems for hundreds of farmers in Butler and surrounding counties. He used an optical level mounted on a tripod to designate the depth at which drain lines should be installed, marking the elevation of each ditch with notches carved on tobacco stakes. Farmers would then stretch a string from notch to notch to serve as a depth guide when setting the tiles in the ditches they dug. They used a wooden jig, called a "preacher" (because it won't lie), to transfer elevations from the string to the bottom of the ditch. He laid out many tile systems while standing in water and told me, "Land is like a sponge and tile pulls it out. Tile it out and you'll get rid of the water." Booster earned a reputation for being the best in twelve counties for his ability to lay out tile drainage systems.

Booster explained how the SCS helped farmers create the outlets that were necessary for draining farms. They would use a dozer to cut a deep trench through the riverbank, often digging down 10 feet or more. He said that the dozer would make a huge hole in the ground that could take days to excavate. The dozer would place a 3:1 slope on either side of the cut, with the bottom being 12 feet wide and the top at least 72 feet wide. He then placed a motorized trencher in the bottom of the ditch, so it could dig down another 6 feet. They unraveled plastic drain line in the bottom of the ditch, surrounding the deepest portions of plastic pipe with steel pipe to prevent crushing, and then covered it with 6 feet of soil. It was common to empty an entire farm's drainage system into one of these deep, artificial outlets.

Booster remembered using a laser level for the first time in 1975 to lay out a drainage system: "We used the laser to mark 10,000 feet of drain line on a farm. Later we found out that something went wrong with the laser head and that all of the plastic drain lines had been buried perfectly level. We left them in the ground anyway, and they're still working fine today."

I asked Booster to tell me why he went to so much trouble to drain wetlands. He replied that a farmer was lucky to get 35 bushels of corn per acre from the swampy ground along the Green River, and these areas regularly drown so that they really only produce in drought years. After tiling and fertilizing, a farmer can now harvest up to 150 bushels an acre every year from the same piece of ground. "Tiling would more than double the worth of the land," he said. When I explained to him that some government programs are now working to return these areas to wetlands, he said, "Why would you mess up land that will grow 150 bushels an acre at $2.00 a bushel?"

Philip Annis and Clay Tiles

Philip Annis grew up on a 400-acre farm in western Kentucky near the Green River in Butler County. His third cousin was James "Booster" Flowers, who lived on a neighboring farm. Philip said, "Each year my grandfather marked the places where crops drown out, then returned to them in the winter to drain them with ditches and tile. He was always fighting drainage problems." In 1953, when Philip was only 8 years old, he remembered riding with his grandfather James Freeman Annis in an old truck to haul clay drain tile out to what was once an oxbow wetland. His grandfather was working to replace a plugged drain line in the oxbow. He watched as workers used shovels to dig ditches from 2 to 4 feet deep, shaping

the bottom of each ditch with a crum tool. An especially deep ditch had to be dug through the riverbank in order to move water out of the oxbow. They used mostly 4-inch-diameter tiles to drain the wetland, 6-inch-diameter tiles for laterals, and 12-inch-diameter tiles for the main line. His grandfather used string as a guide for leveling and poured water in the ditch to see if it had enough slope before the tiles were covered. Water flowed from the main tile outlet into the river year-round, and it was a favorite spot for Phil and his friends to catch catfish. He said that "water coming out of the line was clear, and my friends and I even drank it."[15]

One unusual thing Phil remembered about the large bottomland fields along the Green River was that after spring floods, an occasional large sinkhole, about the size of a car, could be found in a field. This was caused by a tremendous volume of surface water rushing down into a broken tile or a crack left between tiles.

Phil says that it was important that clay tiles touched each other when they were placed in a ditch. This need stuck in his mind as he remembered one of his grandfather's hired hands was convinced that a gap had to be left between tiles so that water could enter them. The worker had buried a number of tiles this way without the knowledge of his grandfather. Over the winter, little sink holes formed over each tile joint, and soil then plugged the line. Each tile had to be dug out, cleaned, and placed back in the ditch.

Phil remembers dozers being used to clear trees from a slough once found on his grandfather's farm. In one case, they were used to clear trees from a 3-acre slough in the middle of an 80-acre field. "One of the trees they removed was huge. Dozers filled in the slough and covered up all the fish that were flopping about. They then installed tiles so that the area would drain," he said.

Locating Buried Drain Lines

Should you ever wonder how long clay drain tiles can last in a field, I suggest visiting Utility

Pipe and Supply Company in Farmers, Kentucky. When you walk in the door, ask James why they keep a clay drain tile on one of the shelves. He will inform you that farmers come by on a regular basis to ask if they have any of them for sale, as they've broken a line while subsoiling a field. He tells the farmer that clay tiles are no longer available and then shows how the plastic fittings they sell can be used to repair the old clay tile line to keep it working.

Detecting the presence of drain lines can be difficult, as physical signs of their presence are well masked. Farmers took great effort to keep soils from settling over buried drain lines, an action that prolongs the life of the systems.[16] Occasionally, drawings of government-assisted drainage projects can be found at local NRCS offices; however, considering that the majority of subsurface drainage was completed without such aid, these records are far from complete.

Two main strategies have been used to install drain lines over the years. The first, called the random system, buries lines in a way to drain scattered wet places that are somewhat isolated from each other; the second is the parallel line system where drain lines are placed in a herringbone or gridiron pattern over an entire field. One typically finds the random system being used in hilly country and the parallel line system on poorly drained areas with little slope.[17]

Clay drain tiles were often buried at considerable depth in hand-dug ditches. Klippart recommended placing clay tiles at a depth of 4 feet, and gave one example where they were placed 14 feet deep.[18] In describing how deep he buried clay tiles, French stated: "Three foot drains will produce striking results on almost any wet lands, but four foot drains will be more secure and durable, will give wider feeding-grounds to the roots, better filter the percolating water, warm and dry the land earlier in the Spring, furnish a larger reservoir for heavy rains, and, indeed, more effectually perform every office of drains." He also said, "It is a well known fact in draining, that the deepest drain flows first and longest."[19] George Waring obtained good results by burying clay tiles 4 feet

deep.[20] And J. Talcott buried clay tile an average of 3 feet deep, extending down 4 to 6 feet when passing beneath knolls near Rome in Oneida County, New York.[21]

"The deeper you put it, the further it will pull," says Richard Bond Jr., who installed field drainage systems in eastern Kentucky and southern Ohio all his life.[22] "I'd get it 30 inches or more in the ground." Richard claimed that drain lines buried that depth would pull water from the surface for a horizontal distance of 50 feet from either side of the pipe. He often buried them much deeper when going after deep holes of water. "The deepest I went was 11 feet in Greenup County," he said. George found that most of the older style clay drain tiles used in the 1800s were buried about 2 feet deep, while the newer clay tiles set since the 1950s were placed at a depth of about 3 feet.

Bond showed me many places where he had installed clay and plastic drain lines over his 50-year career in drainage in eastern Kentucky. All the areas we viewed appeared to be well-drained higher ground—the last place natural resource managers would choose for wetland restoration. He often installed clay tile in fields less than one-half acre in size, on ridgetops as well as near creeks. It was not unusual for landowners to ask him to get rid of wet places where the job would require only 100 feet of tile or less.[23]

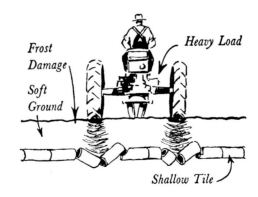

Farmers were encouraged to install drain lines deep enough to keep them from being damaged by frost and heavy modern machinery. (From Keith H. Beauchamp, "Tile Drainage: Its Installation and Upkeep," in *Water: USDA Yearbook of Agriculture* [Washington, D.C.: U.S. Government Printing Office, 1955], 517)

Chamberlain recommended that main drain lines "should follow the 'dry brooks'; that is, take the general direction and location which the water takes, in a wet time, to get off the land; for water, taking its own course along the surface, naturally takes what scientists call 'the path of least resistance.'"[23] He goes on to say that improvement should be made in placing the main line by straightening curves and correcting the grades to make them uniform. Klippart said that the main drain should be located on the lowest part of the farm, and that it should empty into an open ditch from 4 to 6 feet deep.[24]

When discussing where to lay out clay tile drains, Klippart stated: "In the drainage of swamps, or small basin-like depressions, it is customary to cut a main drain through the center, at a depth sufficient effectually to drain the lowest point." To drain large basin-shaped wetlands, he first identified an outlet that was deeper than the lowest place in the basin, then ran a main drain line from the outlet to the deepest part of the basin, where he dug a pit or well. He next buried smaller-diameter drain lines straight up and down the slope in a spoke-like pattern with each emptying into the well.[25]

I asked George McClure to describe how he would go about locating the main line and outlet for a buried drain line system. He said to visualize the path water would follow over the surface of a field should there be a torrential rain. At the point where water leaves the field is a good place to begin looking for the main line. The main line will lead to the outlet, generally following the shortest distance from the lowest edge of the field. The outlet is usually found where the installer could find the greatest drop, which is often an open ditch, road ditch, stream, or river.[26]

Sometimes McClure is able to find clay tiles with a soil probe, providing the soils are not rocky and the tiles are buried shallow. He uses a backhoe to find deeply buried clay tiles or plastic drain lines. George showed me several hayfields in Lawrence County, Ohio, where you could tell where drain lines were buried by a slightly darker shade of green grass growing over them. I also noticed that bulrushes failed to grow anywhere near the locations of fields drained with buried lines.

Small vertical holes on the ground surface often indicate the presence of buried drain lines in an area. While walking over fields and woodlands to design wetland projects, I have found round holes that average 12 inches in diameter above clay tile lines in Kentucky and Ohio, and over a 4-inch-diameter cement drain line in British Columbia. Called "suck holes" by John Haswell, they show where soil from the surface has found a way into the buried drain line through a joint between two tiles.[27]

CHAPTER SEVEN

Massive Machines and Plastic Drain Lines

Ed Stevens owns a small farm in Carter County along the Stinson Road in the hills of eastern Kentucky. His largest field, at 3 acres, was suitable for hay but too wet for row crops. After retiring, he wanted to raise corn for sale at the farmers' market. "The field held puddles of water," he said, so he asked Jimmy Lyons, district conservationist for the NRCS, to help him dry the land. Jimmy was surprised at his request as there had been little interest in draining lands in the last few years. "You can buy good ground cheaper than you can drain it these days," he says.[1]

As Jimmy walked the field with Stevens, he observed a number of crayfish burrows with water near their surface, indicating a high water table. Checking the maps, he found that the silt loam soils were not listed as being hydric, and neither was the field classified as "PC" for "prior wetland conversion." He proceeded to design and lay out a drainage system that involved installing 3,000 feet of 4-inch-diameter slotted plastic pipe at a depth of 2 to 3 feet. The drain lines paralleled each other with 50-foot spacing, with a longer main line lead to the creek to serve as the outlet.

Jimmy marked the locations for the drain lines with tobacco stakes and used a level to measure grade, cutting notches in the stakes for Stevens to use as a guide when burying the pipe. He marked the stakes so that the drain lines would have a drop of 4 to 6 inches for every 100 feet in order to keep them washed clean. Since no government cost-share dollars were available for drain-

age actions, it was up to Stevens to pay for the project.

In October 2004, Stevens hired a backhoe operator to dig the ditches for the drain lines. To measure depth, they stretched a string from notch to notch on each tobacco stake and used a wooden jig to transfer elevation from the string to the bottom of the ditch. He said, "You wouldn't believe how much water flowed out the drainpipe the first couple of days after it was installed. The pipe took care of water on the surface; and I know I'll be able to plow it next spring." Stevens mounded soil over the recently buried pipes to allow for settling. "The dirt has settled quite a bit already. I'll plow and level it up so you don't see the rows this spring." Obviously pleased with how things turned out, he exclaimed, "Those drain lines are going to keep working long after I'm gone!"

Richard Bond and His Tiling Machine

Richard Bond worked hard and was willing to take chances as a drainage contractor. He saw that the government was accelerating its cost-share program for farm drainage systems and decided to purchase an automatic tiling machine. Landowners were lining up to take advantage of an SCS program that paid 80 percent of the cost for approved drainage projects. In 1970, Bond bought a Buckeye Super D Series tiling machine for $18,000. The machine was powered by a gas-

A recently installed system of buried drain lines marked by rows of soil on the Stevens farm in Carter County, Kentucky, after a standard USDA Natural Resources Conservation Service design. A diversion ditch was also excavated at the base of the hill, starting at the utility pole, to help remove runoff.

oline engine with tracks and wheels for maneuvering through wet areas. Because it was 30 feet long, he transported it on a 1.5-ton truck with a tilt bed. Using sights like those on a rifle, he could dig a ditch on grade, cut a groove for the drain

Richard Bond Jr. with his tile machine, installing over 9,000 feet of plastic drain line on the William L. Dailey farm in 1977 along the Clearfork Road in Rowan County, Kentucky, within the Daniel Boone National Forest. (USDA Natural Resources Conservation Service photograph)

line, pack the bottom of the ditch, and install the plastic pipe all at the same time. SCS technicians would design the drainage systems, set the grade, and establish cuts needed for every 100 feet of drain line to be buried. Richard would then place metal stakes at 100-foot intervals along each marked line, which he would sight on to raise or lower the machine for the appropriate cut.[2]

Richard kept busy operating the machine in fifteen Kentucky and Ohio counties for 8 years, with most of the drainage jobs he finished being cost-shared by the SCS. He could install an amazing 11 feet of drain line per minute at an average depth of 36 inches, and the machine could be set to dig down 5.5 feet if it was needed to pass through a rise in a field. He still used his backhoe to dig deeper ditches for outlets or to pass beneath high humps in a field. The machine did not cover the pipe, so he used a backhoe, tractor, or dozer to fill in the ditches.

Richard could lay more feet of plastic drain line in a day, and at a lower cost, than his competition that still used a backhoe or tractor-pulled trencher. While he could hope to install from 800 to 1,000 feet of line a day with the backhoe and two people helping him, with the tiling machine he could install more than 3,000 feet a day alone. He charged farmers from 75¢ to 80¢ a foot to install the 4-inch-diameter plastic drain line, and that price included the drain line that cost him 18¢ a foot. Richard beamed, "I'd start at daylight, go to Vanceburg, and pick up 3,000 feet of tile. That's all I could haul. I'd drive to the job, set the tile, and be home by 2:00 P.M. with over $2,000 in my pocket." He remembers one especially large project in 1975 when he installed 22,000 feet of drain line in one field on the Chuck Hart Farm in Bath County, Kentucky. By the late 1980s, the demand for installing field drainage systems dropped greatly with the cessation of government-sponsored programs. The drainage jobs became smaller and were funded solely by the landowner, so he parked the tiling machine up the hollow behind the barn and brought the backhoe back into service.[3]

Backhoe Installation

A great many drain lines have been installed with a backhoe over the years. Backhoes were readily available in rural areas, and contractors could operate them at a reasonable price. I never realized how much they were used to drain wetlands until I began hiring contractors to help me build wetlands with backhoes. Because much of their work involved installing and repairing buried drain lines, they were quick to grasp the steps needed to reverse previous drainage actions. I also discovered that it was practically impossible to get stuck in a backhoe, as skilled operators can walk one out of the deepest quicksands and mud. Perhaps it was because I had asked so many questions about drainage that late one Friday evening I received a call from John D. Smith of Menifee County, Kentucky, saying that if I wanted to photograph someone installing drain lines I could come over and help him the next day. The forecast was for the thermometer to reach 95 degrees with over 90 percent humidity, so conditions sounded harsh for this native Minnesotan to spend a day working under the sun surrounded by that southern humidity! Nevertheless, to avoid what would be certain ridicule for having an excuse, I agreed to assist, as shown by the accompanying photographs.

George McClure and His Speicher Model 600

George McClure kept very busy draining lands in three Kentucky and eight Ohio counties from 1972 to the late 1980s. He testified that President Jimmy Carter was true to his pledge to "Grow More in '74" and made abundant cost-share dollars available to farmers for drainage projects. At one time, George was one of five drain line installers in the area. In 2005, he was the only one who remained in business.[4]

McClure is a survivor; he works hard, pleases his customers, and is always forthcoming with more efficient ways of doing business. At 62 years

(top) John D. Smith digging a ditch for a drain line on his farm in the community of Sudith, Kentucky. The new line was needed to replace a clay tile line his father had buried by hand 64 years earlier. Trucks operating in the field under wet fall conditions had crushed the old clay line.

(bottom) Wooden jig being used to check the depth of the ditch.

of age, he doubts he will ever retire. He began a drainage business by installing drain lines in 1972 with a Henson wheel trencher pulled by a Forbson-Major front-wheel-drive tractor. Having to maintain and haul two pieces of equipment was a bit troublesome, so he purchased a 1970 Model 600 Speicher trencher in 1975. He remembers paying

Hand-shoveling to fill in low spots and shave off high places left by the backhoe in the ditch.

Anne Smith, John's wife, placing the 4-inch-diameter plastic drain line in the ditch.

John spreading a layer of straw over the drainpipe to help keep fine soils out.

Closing the upper end before the drainpipe is buried.

Final ditch with drainpipe installed before it is covered with soil.

Filling the ditch, leaving soil mounded over the top to allow for settling.

Completed drain line that will be leveled with a tractor and disked the following spring.

$30,000 for the machine, a large sum considering one could buy a new car for $3,000 at that time. The machine is monumental. The cutting wheel stands far above a tall person's head. The trencher is capable of setting plastic drain lines of all diameters in ditches up to 6 feet deep. George claims his best day working with the machine was when he buried 7,000 feet of drain line on a job in Meggs County, Ohio, up the river from Racine.

Plastic drain line being placed with the help of laser-guided equipment on the Russell Hatton farm near the Licking River in Bath County, Kentucky. Use of the machine and laser was a major innovation for its time in April 1981, and the project was featured in *Southern Farmer*. (Photograph by Ben Allen Sharp, USDA Natural Resources Conservation Service)

In a typical day George can haul over 6,000 feet of drain pipe to the job site on a trailer, stretch out the drain lines with a small dozer, dig the trench and place the drain line with the trencher, and then cover the drain line with the dozer. He uses a backhoe that he mounted to the back of a dozer to dig open ditches, excavate holes to join pipe sections, and dig outlets more than 6 feet deep.

In the mid-1970s, George spent $10,000 to retrofit the trencher so that the ditches it dug were automatically controlled by a laser. The laser eliminated the labor needed to set targets or a survey hub in a field. The greatest benefit of using a laser-controlled system was that it permitted him to accurately set buried drain lines in level areas with as little drop as 0.1 feet per 100 feet, allowing for the draining of level areas at a reasonable cost.

Fuel prices soared in the 1980s, so George looked for ways to increase efficiency and came up with a modification of the Speicher so that it would cover the drain pipe as it moved forward. He said that this change "really helped the farmer. I'd always make such a mess going back and forth with the dozer to cover the lines. By making the trencher cover the pipe, all I had to do was some finish grading with the dozer, and I could do this by going forward in high gear."

To help drain those very wet areas, he maintains a 300-foot-long winch with three speeds on the front of the trencher. I watched him use the winch to pull the trencher through mud and standing water as he installed a 4-inch plastic drain line in an area disturbed by gas line construction in Gallia County, Ohio.

After spending a day with George driving around southern Ohio, I was offered an overview of his successful business by looking at a large number of places he helped drain. I discovered that he often works to improve drainage systems on lands, not to replace them. In many areas, he worked to locate existing buried clay drain lines, so that he could place new plastic lines between them. He generally spaces new drain lines at 60-foot intervals but will tighten up to only 40 feet in extremely wet areas.

To start a job, George typically constructs open ditches to remove standing water. If they don't dry an area enough for farming, he returns to bury plastic drain lines. He finds that drain lines are more likely to work as planned if they are set on ground other than quicksand, and that open ditches will help firm soils for the installation of buried drain lines. George claims that sand tiles are less likely to plug when used in saturated sandy ground with flowing water. Sand tile is plastic drain pipe with slits cut so narrow they can barely be seen.

While George and I were traveling around southern Ohio examining drainage projects, he regularly asked me about government programs for wetland restoration and about the techniques used to build wetlands. He wondered if it would be at all possible for the government to begin a program to certify contractors such as himself in wetland restoration, as this would assure landowners they were hiring a qualified wetland builder, an expert in the field. It dawned on me that with his keen interest and willingness to venture into new kinds of work, who would be better to hire than George to begin restoring many of the drained wetlands in Ohio?

Surface Inlets

Surface inlets were often combined with buried drainage systems to help remove standing water from wetlands. A surface inlet provided a direct route for runoff to rapidly enter a system of buried clay tiles. Colonel George Waring provided a detailed account of how he used clay tiles, a surface inlet, and a sediment basin to successfully drain a swamp. In summary, he dug a deep ditch along the upper edge of a field at the base of a rock hill to collect runoff. The opening and upper end of a tile line were placed in the bottom of the ditch, and the ditch was filled with stone to the surface. Runoff traveled off the rock hill and down into the rock-filled ditch directly into the clay tile line. The clay tiles then carried water downhill to the lowest place in the swamp, where

George McClure uses his Speicher Model 600 to install plastic drain line on private land in Gallia County within the Wayne National Forest in Ohio on November 2, 2003.

a tile basin or sediment trap was dug. Drain lines were then buried at 20-foot intervals through the middle of the swamp. Water from the drain lines was collected in the tile basin, where a larger tile line carried water beneath the creek bank into the creek. The sediment trap had a removable top for inspecting the openings of the lateral and main drain lines and for removing any soil.[5]

The benefit of introducing air directly into a drain line was noted by I. Whiting in 1855: "I am of the opinion that all tile drains will discharge their waters more freely by having air admitted into the upper end, that is where drains are long and put down 3 or 3.5 feet, in a close, retentive soil."[6]

George McClure's Speicher Model 600 laser-guided trencher buries a large-diameter plastic drain line in a wet field in Ohio during the early 1970s.

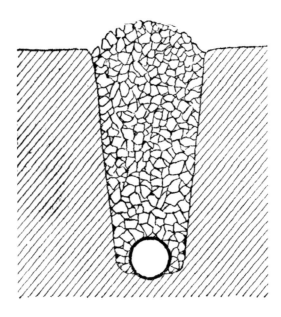

Drawing from 1919 showing how to construct a surface inlet of rock over a buried clay tile line to rapidly remove standing water from a wetland. (From Charles G. Elliott, *Engineering for Land Drainage: A Manual for the Reclamation of Lands Injured by Water*, 3rd edition [London: Wiley, 1919], 152)

Before Drainage

Stream

After Drainage

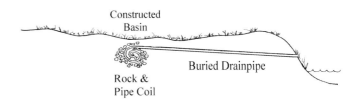

Constructed Basin

Buried Drainpipe

Rock & Pipe Coil

Surface inlet construction with plastic drain pipe. (Drawing by Dee Biebighauser)

Charles Elliott recommended placing surface inlets in depressions where water accumulates. He claimed that surface inlets accelerate water velocity in a buried tile system, thereby increasing its effectiveness. The type of surface inlet he commonly recommended was a hand-dug section of trench, 3 to 6 feet long, filled with broken stone 3 to 4 inches in diameter. A tile line with open joints, or even a T, was placed in the bottom of the ditch to serve as an inlet. The stones in the trench also helped filter water and reduced the chance the inlet would plug. Elliott also described how surface inlets could be combined with silt basins to prolong the life of a drainage system.[7]

Hickenbottom Surface Inlet

A Hickenbottom is a type of surface inlet that is placed in the bottom of a wetland or depression to rapidly drain surface water into a system of buried drain lines. The Hickenbottom consists of an intake riser pipe with holes in it; it is constructed of plastic or steel and extends vertically above the ground. It is used to quickly remove standing water from both large and small depressions before crops are damaged. The Hickenbottom catch basin became popular in the late 1960s with the development of PVC pipe and fittings. Early risers were constructed of PVC pipe with holes drilled by hand. The factory-built orange-colored Hickenbottom is most commonly seen in fields today.

The presence of a Hickenbottom surface inlet provides strong evidence that an emergent wetland was once present at the site. "When you see a Hickenbottom, you can bet that the closest creek bank was higher than the field, and that the field once held pockets of surface water," says Jimmy Lyons, district conservationist with the NRCS.[8]

The accompanying photographs show the construction of a grassy waterway and a Hickenbottom surface inlet with the SCS providing technical guidance.

(top) Wetland drained with a Hickenbottom surface inlet one day after a 3-inch rain near the Licking River in Rowan County, Kentucky.

(middle) Same wetland 2 days after the rain.

(bottom) Same wetland 3 days after the rain.

Vertical Drains

Perhaps one of the most ingenious ways for draining what we now call isolated wetlands was by the use of a vertical drain.[9] A hole, usually 6 inches in diameter, was hand-dug with an auger or drilled with power equipment through fine-textured soil layers that formed the wetland down into permeable layers of sand, gravel, or even a cave. Gravel, rock, or cinders were then placed over the entrance of the hole to keep it from plugging.[10] Such vertical drains were capable of rapidly and reliably removing waters from the smallest to the largest wetlands.

John Johnstone in his 1808 book about draining describes how boring in the bottom of ditches was used to discharge water from wetlands and lakes down into deeper layers of permeable soil. He recounts that when Dr. Nugent traveled to Germany in 1766 he observed and described the following method for draining marshes with no available outlet: "A pit is dug in the deepest part of the moor, till they come below the obstructing clay, and meet with such a spongy stratum as, in all appearance, will be sufficient to imbibe the moisture of the marsh above it." Soil-covered stone drains were made to discharge water into the pit, which was protected with flat stones and covered with earth.[11]

John Klippart described how wells and open drains were used to drain a "perfect morass" between Canton and Massillon in Stark County, Ohio, around 1818. Wells were sunk as deep as 66 feet to penetrate sand, gravel, and a hardpan of blue clay in order to create usable land.[12]

Instructions on how to install vertical drains were provided by Ayers and Scoates in their book about land drainage. They described how vertical drains had been used to drain wetlands in Iowa and Wisconsin. Extensive networks of buried drain lines were installed to carry water into these hidden drains. The vertical drain allowed farmers to eliminate wetlands where no gravity outlet existed for removing surface or subsurface waters. In some cases, individuals were able to use a hand auger to install a drain by penetrating the

(upper left) Using a dozer to excavate an open-ditch, grassy waterway with gradually sloping sides leading down to the creek on the Chuck Hart farm in Bath County, Kentucky, in 1991. (USDA Natural Resources Conservation Service photograph)

(left) Placing 4-inch-diameter black plastic drainpipe in a deep ditch made with a trencher in the bottom of the dozer-dug ditch. The presence of two levels shows that great care is being taken to keep slopes gradual for preventing erosion. NRCS employee Jimmy Lyons is the center person in the photograph. The section of pipe in the foreground will be used to make the Hickenbottom surface inlet. (USDA Natural Resources Conservation Service photograph)

(below left) The Hickenbottom surface inlet riser and graveled catch basin used to prevent water from standing at the base of the hill and in the finished grassy waterway. (USDA Natural Resources Conservation Service photograph)

(top) Close-up of Hickenbottom surface inlet riser with holes drilled by hand. (USDA Natural Resources Conservation Service photograph)

clay in the deepest part of the wetland to carry water down into a sand layer. Six-inch-diameter clay tiles were often used as casing for these shallow holes, with iron being used to line deeper wells. Examples were given from northern Iowa, where a number of vertical drainage wells had been known to function for over 30 years, which meant that this practice has been in use since the late 1800s. They described how one vertical well in northern Iowa was used to drain 320 acres of wetland down into a limestone cavern.[13]

From the accounts in the literature, it appears that the vertical well method for draining wetlands was very effective and was possibly used extensively in the United States. Unfortunately, I have not been able to locate an actual wetland eliminated by the practice. I believe that wetlands drained by this method would be challenging to restore because it would be difficult to locate the small inlet for the buried well. Therefore, locating a dry basin of fine-textured soils that is distant from a ditch, stream, or drop-off should alert the land manager to the possibility that this technique may have been used to drain the area.

Wetland Drainage and the Water Table

Wetland drainage could have quite an effect on the elevation of groundwater in a community. Early authors told stories of how wells could go dry following the drainage of bogs and marshlands in an area. For example, in order to drain lands for farming in the late 1700s, Joseph Elkington tapped and diverted a spring to a depth of 16 feet lower than the bed of a nearby small river. His actions were responsible for drying all the wells in the neighboring town of Lutterworth, Scotland.[14]

Henry French told a story about how, before 1860, draining a swamp in New Hampshire with clay tiles may have affected the elevation of the water table and his neighbors' wells:

In our good town of Exeter, there seems to be a general impression on one street, that the drainage of a swamp, formerly owned by the author, has drawn down the wells on that street, situated many rods distant from the drains. Those wells are upon a sandy plain, with underlying clay, and into it, and may possibly draw off the water a foot or two lower through the whole village if we can regard the water line running through it as the surface of a pond, and the swamp as the dam across its outlet.[15]

Edmund Ruffin Esq. gave an example of drying up a well a half-mile distant from where wetlands were drained in southern Virginia before 1857. Saturated beds of sand were positioned from 4 to 8 feet below the surface of swamps in his district. Using the drainage system developed by Elkington in the 1700s, Ruffin dug ditches at wide intervals through lowlands and then bored down into the pressurized aquifer with an auger to release and direct waters into a system of buried drain lines, successfully converting wetland into farmland.[16]

Early land drainers were determined to find ways to remove surface water and groundwater, so it is not unreasonable to believe that their practices affected the water table on neighboring farms. Henry French expounded on the position that it should be the legal right of a landowner to drain his own field, regardless of how it would reduce waters available to farmers and businesses living downhill from the deed. Mill operators disagreed, stating that the consistent flow of streams was essential to their operations. Farmers who lived downhill from drainage actions were concerned about losing the water they needed for livestock and irrigation and with the possibility that their land would become dry and worthless.[17]

Filling and Leveling

Farmers took considerable effort to fill low areas that held water on their land. John Johnstone recommended leveling the surface of a bog or marsh after it had been completely drained.[1] A. I. Root described the importance of filling low areas where water may stand by taking soft earth from higher portions and throwing it into depressions to prevent harm to crops. He stated, "I recommend having ground gradually brought in shape so that there will be no depressions where water may stand for even an hour." He also encouraged farmers to shape the land so that water would flow off in some direction.[2]

Small shallow wetlands were filled by repeated plowing to move soil from higher ground, thereby covering them over time. Tractors with blades were also used for filling small wetlands and for leveling fields. Dozers and scrapers were later engaged to complete major field-leveling projects. Many landowners went to this expense out of a desire to have clean, completely productive farms, where the economics of filling small wet places was not always the motivation for their actions.

The practice of land leveling has been used by farmers for generations to fill small wetlands in fields. "This practice fills the pothole with soil, leaving a level field" according to Robert W. Burwell and Lawson G. Sugden: "A farmer who plows around a wetland in one direction can actually plow his low hills into a basin and fill it over a few years. Generally, that is done only on small, temporary potholes."[3]

In 1988, Jimmy Lyons began working for the NRCS in eastern Kentucky, when over 75 percent of his job involved helping farmers with drainage problems. Jimmy said that fields with water problems often had a number of low areas that held standing water. An effective way to dry these water pockets involved using heavy equipment to reshape the area so that runoff would collect in a low place near the middle of the field. A sump would be excavated in the depression, and a ditch dug from the sump to an outlet, such as a creek or roadside ditch. A 4- or 6-inch-diameter plastic drain line would then be buried in the ditch between the sump and the outlet. A coil of plastic drain line was used as an inlet in the sump, and the sump would then be filled with larger rock, such as #2 limestone. The reshaped field and drainage structure would then act like a funnel to remove water from the field. He said that you could use a Hickenbottom surface inlet instead of a coil of plastic pipe; however, these were more expensive. Sometimes the gravel over the inlet would be covered with geo-textile fabric and soil and planted to sod. Farming operations could then continue over the entire drainage system with no loss of land. He said that this type of drainage practice also worked great for carrying water away from springs that surfaced in a field.[4]

Richard Bond also installed a number of drainage systems like these that involved the use of buried coils of plastic drain line in eastern Kentucky. He liked to leave two circles of pipe, or about 15 feet, in the bottom of the sump to drain a spring and found that the system also was quite effective for taking care of those "big holes of water."[5]

In the early 1980s, contractor Earl J. Osborne of West Liberty, Kentucky, was asked to visit with Bee and Joanne May at their farm in Morgan

County along Grassy Creek. The Mays showed Earl an oxbow of Grassy Creek filled with standing water and flanked by huge, 36-inch-diameter bottomland hardwood trees. The Mays were not able to farm the 2-acre oxbow and the land near the oxbow because the ground was too soft; Earl said it was "nothing but a swamp." The couple listened to Earl's suggestions for improving the site, liked what he had to say, and hired him on the spot.[6]

To begin the job, Earl proceeded to cut a deep channel along the base of the mountain, stockpiling large quantities of soil. He redirected Grassy Creek into the straight channel, cutting off water from the bend in the creek. Felling numerous huge trees with a chainsaw, he pushed them into the oxbow with the dozer. The large piles of dirt from the excavated creek channel were used to cover the bend in Grassy Creek and its associated wetlands with up to 7 feet of soil. The job cost thousands of dollars and was paid in full by the Mays without government assistance. It is interesting to note that this work was done well after Section 404 of the Clean Water Act was enacted in 1977. With hundreds of cars traveling by each day along State Highway 203, apparently no one ever phoned the Commonwealth of Kentucky or the Army Corps of Engineers to see if a permit had been issued for filling a section of Grassy Creek.

More than 20 years later, the Mays's mowed hayfield looks quite natural as you drive by on Highway 203. If you pull over to the side of the road and look down carefully, there are some signs that indicate the area was changed; the first is a small dark spot with a glint of water in the field. Earl says that water standing in the low place was caused by soil settling over the deepest part of the filled oxbow. Looking closer, you can also see a slight dip in the natural levee along Grassy Creek that marks where the old and new creek channels were merged downstream of the filled wetlands.

Conversion to Deep Ponds

Many wetlands or parts of wetlands have been converted to deep ponds for watering livestock

Ephemeral wetland being filled with a tractor and blade on October 10, 2003, in Morgan County, Kentucky. More than 2 feet of soil was moved from a ridge in the field into the wetland.

A number of small wetlands up to 4 feet deep in this 12-acre field were filled and leveled by Wayne Williams in Jackson County, Ohio. In the 1980s, diversion ditches were constructed to divert runoff and to remove surface water.

and fishing. Generally, the only way to discover if a natural wetland has been deepened is to examine old aerial photographs or question the landowner or the contractor who performed the work.

Dozer operator Earl J. Osborne built thousands of ponds for farmers on private lands in eastern Kentucky from 1970 to 1990. He said that when he showed up for work, the landowner would al-

Small scrapers like this one owned by George McClure were used to cut ditches, fill wetlands, and level ground.

ready have a good idea of where the pond should be located for watering livestock. Earl built many of these deepwater ponds on locations that had shallow water, gray-colored soils, and aquatic plants such as bulrushes and willows growing in them. He said that landowners were reasonably sure that these areas could be made into a deeper pond since they already held some water. One of

Historic oxbow wetland and creek channel along Grassy Creek in Morgan County, Kentucky, which had been filled with 7 feet of soil in the early 1980s on the Mays farm by Earl J. Osborne.

the first steps he would take in construction was to dig a ditch to drain standing water from these sites.[7]

Most of the farm ponds Earl built were from 16 to 20 feet deep. Constructing these ponds required huge quantities of soil to build up a high dam, which resulted in large areas being scraped during construction, generally removing any evidence of a natural wetland that may have been present on the site.

Personnel from the NRCS often assisted farmers with these projects by testing soils, designing the pond, and helping oversee its construction. The NRCS pond construction program was very popular with landowners, as it reimbursed them for the majority of costs associated with building these needed watering sites. In my discussions with them, retired NRCS district conservationists confirm that areas we now call ephemeral wetlands and wet meadows were favored places to build farm ponds from 1940 to 1990. Landowners did not consider these swampy places to be wetlands and regarded such areas to be of low value, as they were difficult to farm.

Loren Smith reported that a common form of hydrological modification to playa wetlands in the Great Plains of the United States was pit excavation. He found that many of the larger playas had pits dug in them to aid in the irrigation of crops grown in the watershed surrounding the wetlands.[8]

Senior technicians and contractors at the Daniel Boone National Forest have told me that beginning in the 1940s forested ephemeral wetlands were routinely selected as locations to build deeper ponds in order to provide water for firefighting and the once uncommon white-tailed deer. They often waited until these areas dried in the fall before digging them out to a much greater depth.[9]

In 1988, when I began working with personnel within the Daniel Boone National Forest to locate sites for building wetlands, I was regularly taken to small areas that were already wetlands. Even though these forested, wet meadow, and ephemeral wetlands were uncommon features on the

Permanent water pond created during the 1960s in a natural ephemeral wetland within the Daniel Boone National Forest in McCreary County, Kentucky. Saplings on the left are growing on soil piled by the dozer.

landscape, they were favored because they already held some water and it was believed that with excavation they were capable of holding even more water. Both employees and contractors placed little value on these shallow-water ecosystems, believing that the much deeper waterholes they sought to build were more important.

Over the years I've seen where well-intentioned individuals across North America are often drawn to modifying existing natural wetlands when beginning a restoration program. Unsure of how to select sites for wetland construction, they feel more confident of success when working in a basin that already collects some water and shows that it can support aquatic plants such as cattails and bulrushes. Many have trouble believing that dry, sloped locations lacking aquatic plants can be successfully shaped into wetlands, and often for less money than wetter sites. Working on dry sites can be a positive thing, resulting in the addition of wetlands to the landscape, instead of merely changing one type of wetland to another.

Wetland Drainage Stories

A report prepared by Thomas E. Dahl in 1990 contained now widely used data listing the extent of wetland loss in the United States from the 1780s to the 1980s.[1] Like so many others, I was shocked by the magnitude of wetland destruction the first time I viewed the findings. I wondered where these historic wetlands had been located and whether it might be possible to find traces of them on the ground. This began my quest to find drained wetlands and to identify clues that would indicate their existence.

Randy Smallwood Farm

Randy Smallwood, district conservationist with the NRCS, purchased a 70-acre farm in Menifee County Kentucky in 2002. The farm straddled a narrow ridge within the Daniel Boone National Forest. Long, steeply sloped fields were bordered by vertical sandstone cliffs typical of the Pottsville Escarpment. Randy helped his father plant tobacco on the more level parts of this farm they had once rented from Edward Bryant in the 1990s. His father determinedly worked to plow a low portion of the largest tobacco field each spring, which Randy discovered later to be a drained wetland. The depression used to be an ephemeral wetland supporting at least four large sweet gum trees. More of the depression could be plowed in dry springs than in wet years. The tobacco never did very well in the wet soils around the low area, always growing shorter than that planted on higher ground. Around 1980, Bryant had attempted to drain the wetland, wanting to

plant it to tobacco, so he hired backhoe operator John D. Smith to dig a ditch down the center of the wetland. The ditch cut through a high ridge that had runoff trapped to create an outlet for the water. A 4-inch-diameter, perforated plastic drain pipe was buried in the bottom of the ditch, which was then filled to make it possible to farm over the site. The government did not assist with this drainage project.[2]

All that remains of the wetland today is a 0.25-acre depression surrounded by bright green winter wheat sown to halt erosion on the harvested tobacco field. A shallow ditch runs down the middle of the depression, ending at the edge of the ridge. The ditch likely represents soils that settled over the buried drainpipe. The deepest part of the drained wetland has not been plowed in over 5 years and is growing up to shrubs and trees. A 6-inch-diameter red maple grows where large rocks were dumped on the site. The soils in the unmowed area are saturated, having a sandy-silt loam texture; they are gray-colored with black manganese mottles. We found the drain tile outlet hidden beneath two large rocks on the back side of the ridge, about 200 feet downhill from the drained wetland.

Several points are worth noting about this drained wetland. First, it is located on top of a narrow mountain ridge of sandstone where one would least expect to find a wetland. Second, in spite of its small size, considerable effort was taken to drain the site. The high value placed on level land for raising tobacco made it worth the owner's time and expense to ditch and tile the location. However, the drainage activity was

Hand-dug ditch in which Donald *(left)* and Den *(right)* Smallwood (Randy Smallwood's father and uncle) placed 4-inch clay tiles to improve a tobacco field during the 1950s in Menifee County, Kentucky. (Photograph taken by an employee of the USDA Natural Resources Conservation Service who assisted with the project)

only partially successful, as about one-fourth of the original wetland still remains, although in a modified condition.

Randy offered some observations concerning the drained wetland on his farm: the addition of a couple of more buried lines would have effectively drained the entire site. Another concerned how the contractor, Smith, must have used a handheld level to measure grade for the drainage ditch, possibly resulting in the installation of a pipe that may not have had enough drop at its outlet to keep sediments washed clean. Looking over the small site, Randy commented, "When you consider how much work it took to drain a little place like this one, just think how many changes they made to create those larger fields."

Shrubs and trees denote the deepest part of the drained wetland on the Smallwood farm.

The drain line outlet is located beneath the two large rocks in this washout on the Smallwood farm.

Rand Ragland Farm

Residents of Bibbs County, Alabama, are justifiably proud of the Cahaba River, one of the few rivers in the South that escaped damming by the Army Corps of Engineers. Rand and Cindy Ragland own a 1,200-acre farm along the banks of the Cahaba River, which winds through stands of old cypress on its way to the Gulf of Mexico. Farm fields are large and level on this vast expanse of bottomland. The Raglands most recently have harvested hay from these fields, a major departure from how the farm once produced quantities of corn and soybeans.[3]

Above the river, near the base of a hill along Highway 5, one finds the remains of a 33-acre ephemeral wetland. In the early 1960s, Rand's father took measures to increase production on the farm, which included draining the wetland for agriculture. Dozers and backhoes were used to dig a series of open, parallel ditches to carry surface and subsurface water to the Cahaba River. The drainage ditches were deep, from 3 to 8 feet below the surface. Some led to what appeared to be natural streams bordered by large hardwoods but were really older ditches hand-dug in the 1800s. The wetland drainage project was planned, funded, and carried out by Rand's father, with no assistance from the government.

(top left) Drainage ditch hand-dug in the 1800s on the Ragland farm near the Cahaba River in Bibbs County, Alabama.

(middle left) Most of the drained wetland on the Ragland farm was converted to a field of now-mowed hay.

(bottom left) Longleaf pines planted in the field, once a wetland, were killed when beaver dammed this drainage ditch on the Ragland farm.

Much of the drained wetland is now mowed hayfield, the wettest portions having grown up to trees. Rand planted part of the drained wetland to pines several years ago in a cost-share program sponsored by the Alabama Division of Forestry. However, many of the planted trees died when beaver began damming the ditches. Rand works to keep the ditches open by breaching the dams and by hiring trappers to catch and remove the beaver. He mows the edges of the ditches so they won't grow up to trees and shrubs. He doesn't believe that buried drainage tiles were used to dry the wetland; however, Cindy has found a few clay tiles lying around the farm.

The soils in the hayfields are of a silt-loam texture and are brown in color. The vegetation is dominated by orchard grass and fescue. The presence of straight, open, and deep drainage ditches bisecting nearly level fields are the best physical signs that the area was once a wetland.

John D. Smith Farm

The land that John D. Smith farms in Sudith, Kentucky, could illustrate a beautiful calendar. His fences are straight, fields are green, and weeds are absent. It is hard to believe that wetlands were once common amid these productive meadows of grass, clover, and soybeans in the mountains of eastern Kentucky. John and his late father, Wiley Smith, have fought water on this farm for more than 100 years. They've changed creeks, dug ditches, leveled fields, hauled fill, and buried miles of tile to create a landscape without wetlands. John remembers when six or more shallow marsh, ephemeral, and forested wetlands were located on his and his neighbor's farms. John makes it clear that he never participated in government programs offered over the years to drain these wetlands and that he, his father, and his grandfather designed, funded, and carried out each drainage project on their own.[4]

Location of historic wetlands, streams, and drainage structures on the John D. Smith farm in Menifee County, Kentucky. *Key:* dark blue circle, historic wetland; dark blue ribbon, historic stream; aqua line, open drainage ditch; red line, buried drain line; green spot, spring.

Wetland 1

This was once a forested, 1-acre, spring-fed wetland on a 2 percent slope. The spring near the base of the hill was used to water cattle, as it held a permanent pool of water. The county built a new road in the 1930s, blocking cattle access to the pond, so Wiley Smith drained the wetland by hand-digging a 3-foot-deep, 530-foot-long ditch from the wetland to Salt Lick Creek. He then hand-placed 4-inch-diameter clay drain tiles in the ditch, filling it to original ground level. The line plugged a few years later, so Wiley dug out all the tiles and cleaned mud from them, reburying them one at a time. John remembers his father getting upset with him and yelling as he jumped across the ditch he had dug out for the second time. "It must have been aggravating for him to see me knock dirt back into the hole," said John. In the early 1960s, the clay tiles plugged again, and this time John used a backhoe to dig them out, replacing them with a 4-inch-diameter plastic line. A year later, this line also became blocked, so he replaced it with a larger, 6-inch-diameter plastic drain line. John had trouble finding the 6-inch line locally so drove all the way up to Leba-non, Ohio, to purchase the large coils. The drain line has not plugged since.

John created a simple, yet ingenious, way to keep debris from entering the larger drain pipe. At the start of the line, which had been placed in the origin of the spring, he buried a 5-gallon plastic bucket with holes drilled in it. He cut out a 6-inch-diameter hole in the side of the bucket and placed it over the entrance of the drain line. Waters from the spring now collect in the bucket and feed into the drain line before traveling to the creek. John clears away leaves that can block holes in the bucket once a year to keep them free flowing. A few larger trees have now grown up around the spring.

There are few outward signs showing that there used to be a wetland at this location. No evidence can be seen of the original basin that contained the wetland, as John has filled and leveled it with a tractor and blade. The majority of the original wetland area does not have hydric soils. However, the soils within 20 feet of the spring at the base of the hill are saturated, with iron mottles. And soils immediately around the spring are gleyed with the characteristic gray appearance. A small clump of trees grows around the spring's location, showing that for years the ground has been soft and too wet to mow with a tractor. Perhaps the best clue to the wetland's existence is that the spring only saturates a small patch of land.

Bucket with holes placed over the inlet of the buried drain line in the center of the spring that fed wetland 1 on the John D. Smith farm, to keep it from plugging with soil and debris.

Wetland 2

John Smith tells that when he was a boy, he caught frogs in this wetland. It had been a shallow marsh about 1.3 acres in size that was fed by a small stream and two springs. A few large maple trees grew on the site, and their roots spread out over the surface of the ground. His father, Wiley Smith, moved the stream out of the wetland by hand-digging a new channel. The stream is now in a straight ditch that runs along a fence that also marks the property line. His father used a pair of mules to plow a number of parallel ditches through the area, forming narrow ridges of soil called "lands" between the ditches that were dry enough to produce one cutting of hay each fall. However, many times after a hard winter or a rainy spring, the area was too wet to plow. John recalls that when he was a young boy, water would flow down the furrow behind him as he plowed the field with a pair of mules. He farmed the drained wetland for years, working around all the ditches. In the early 1950s, John grew tired of the open ditches, so he hired a backhoe and operator to dig deeper ditches, and placed 5-inch clay tiles in the bottom of the main ditch and 4-inch clay tiles in two spur ditches to carry water from each spring. The 5-inch line emptied into a ditch along a road the county had built. Over the years, he has plowed the area to throw soil over into the low spots and used a dozer to fill and level the field. Some time later, the buried tile line carrying water from one of the springs became plugged, so John used his dozer to place soil over the spring. The spring is now hidden beneath what appears to be a natural slope along the edge of the field. The wetland is now a gently sloped soybean field that can be plowed in early spring. The only obvious physical sign that indicates the field used to be a wetland is the shallow, straight ditch that runs along the property line. A careful examination will reveal a small patch of sedge about 100 feet down from where the spring was buried, which may indicate where waters from the spring have found a new way to the surface.

John walked alongside me with interest as I took a number of soil samples in the field. We found that the water table was more than 5 feet below the surface. Its silty clay-loam textured soils were brown colored, measuring 10YR 5/3 and 5/4 on the Munsell Soil Chart. I asked John if he remembered what color the soils were when he began plowing the field. He said, "Why, they were as blue as blue can be."

John walked directly to the place where the main tile line emptied into the road ditch. He found the outlet by probing with a shovel until he struck something solid and then used his hands to remove debris that covered the pipe. At first, muddy water flowed into the ditch, but within

View from the center of drained wetland 1 toward the spring, located in the unmowed area between the trees.

seconds, a steady stream of clear water bubbled up from the ground. Considerable water pressure has kept the line from plugging for over 50 years. The tile outlet had become so well hidden that finding it would not have been possible for anyone not involved with its installation or maintenance.

On August 25, 2005, NRCS soil scientist Wes Tuttle from North Carolina taught a class at this

View of drained wetland 2 on the John D. Smith farm.

John D. Smith with the drain tile outlet that removed water from wetland 2.

drained wetland, teaching participants how to use an electromagnetic induction meter (EM 38) and a ground-penetrating radar (GPR) unit. The handheld EM 38 is a conductivity meter that assesses changes in soil properties across the landscape. The EM 38 responds to changes in salts, clay content, type of clay, and moisture in the soil. It is used to find changes in spatial patterns in

Downhill view of wetland 3, more than 60 years after draining. The barn was built over a buried line of clay tiles that still carry water from the former wetland.

soil up to 60 inches below the surface. After walking over the field in a grid-like pattern with the EM 38, Wes produced a color map on his laptop computer that showed darker shading in areas that John D. Smith said basically corresponded to the location of the original wide stream and wetland.[5]

The GPR unit we used consisted of a large box with a trailing antenna, a harness for pulling the box, and a laptop computer for recording data. The GPR sent a low-frequency signal into the earth as it was dragged along the surface of the ground. Wes has found that the GPR works best in drier soils that are not clay. John Smith walked alongside him as he sampled the field with the GPR and observed that it produced what appeared to be a characteristic mark whenever it was pulled over the location of buried clay tile drainage lines.

Wetland 3

John Smith's sister Gertrude and her husband, Clyde Holland, lived on an adjacent farm, and their 1.1-acre forested wetland held pools of water year-round. The wetland could not be farmed until after 1942, when it was drained by the Hollands with technical assistance from the SCS. Clyde began the project by cutting and removing the trees growing on the site. His first drainage action involved digging a ditch at the base of the mountain to prevent runoff from reaching the wetland. Clyde wanted to dig the ditch down the hill and then follow the property line. However, John suggested that he dig the ditch along the base of the hill for a greater distance and then follow the property line so he would not have to cross the ditch with equipment to access the upper end of the field. Clyde listened to John and followed his advice. The second drainage action involved hand-digging a number of ditches in the field, with the main one directly through the middle of the wetland. They buried 4-inch-diameter clay tiles in these ditches.

Around 1947, the Hollands built a new barn

Wetland Drainage, Restoration, and Repair

on their farm. The workers hit a clay tile line in two places when hand-digging the footers for the barn supports. The workers repaired the tile line, and it still passes under the barn today, carrying water from the drained wetland to the creek. I've found that what happened to the Hollands when building their barn is typical for other lands that were drained. People forget where buried drainage structures are located, or they may not even be aware of their presence on the farm.

The ditch that runs along the base of the mountain is the best outward sign that the field with its gentle 1 percent slope was a wetland. The field was plowed and sown to winter wheat when I visited the location, and I found its soils were brown in color with a silt-loam texture. Some small rocks were scattered on the surface, which may have been left over from the filled creek channel.

Wetland 4

This was a 2.2-acre oxbow wetland originally formed by the periodic flooding of Salt Lick Creek. The wetland served as an overflow channel under high-flow conditions and would scour a bit deeper each time the creek flooded. The wetland held water when John's grandfather bought the farm in the early 1900s. Wiley Smith took over management of the farm when he was only 12 years old, after his father passed away.

Wiley Smith hand-dug a ditch through the wetland down to Salt Lick Creek in the early 1930s. He then buried 4-inch-diameter clay tiles in the ditch, effectively draining the wetland. This same tile line was also used to drain part of wetland 2, as described above. Even with the buried tiles, the area was often too wet to plow in the spring, meaning that Smith could not depend on it for raising corn every year. Wiley Smith's death was related in part to his successful drainage of this site. When he was 83 years of age, in the fall while he was cutting and burning brush near the creek, his clothing caught fire. He ran over to the deepest part of the old wetland that generally held water to douse the flames, but it was dry.

Grassy field, once wetland 4 on the John D. Smith farm. The trees in the background are growing along Salt Lick Creek.

Complications from the severe burns he received were responsible for his death.

John began filling the low area that once held the wetland in 1989 by spreading the word that he had a place that needed fill. People brought in loads of sawdust, road construction debris, and even leftover topsoil to his farm. Over the years, John has raised the elevation of the oxbow by at least 2 feet. The wetland now has a slight rise running down the middle and can be plowed in early spring, resulting in consistently high yields of corn. There are no signs that the creek once flowed through the area or that a wetland was present. Soil tests taken on the site show a mixture of textures and colors, none of which had hydric characteristics.

Wetland 5

This 8.9-acre scrub-shrub wetland was located on John Buford Shrout's farm, across the creek from John D. Smith's farm. The area was too wet to plow, and its soils could only be worked after extended drought conditions that occurred infrequently in eastern Kentucky. Previously constructed open ditches had kept water from standing on the surface; however, the soils remained saturated, and the area was covered with alder shrubs.

(top) View of drained wetland 5 on the Shrout farm in Menifee County, Kentucky. The shed was built over a drain line in what was once the scrub-shrub wetland.

(bottom) One of two small ponds built to collect groundwater for draining wetland 5.

Shrout hired the services of Smith and his dozer in the early 1970s for about 3 weeks to clear and drain the wetland. John began the project by digging a ditch along the base of the hill to divert runoff. The ditch now runs along the upper side of an access road. Most would view the ditch as one constructed for the road, yet its original purpose was to help dry the wetland.

John then dug two small ponds near the upper edge of the wetland. He remembers hitting bedrock about 8 feet below the surface in each pond, both of which filled rapidly with groundwater. Shrout brought in a contractor from Sharpsburg, Kentucky, to dig a ditch from each pond down to Salt Lick Creek. John remembers the contractor using an old 8N Ford tractor to pull a ditching machine. John was dissatisfied with the ditchdigger's work: "[The ditches] went up and down and were kind of wavy." John went ahead and buried 4-inch-diameter plastic drain lines in each ditch, covering them with soil. The entrance of each drain line also served as the overflow for each pond, with the ponds acting as a sump to draw in groundwater from the surrounding landscape, capturing sediment and effectively draining the wetland. The project worked so well that Shrout grew corn on the site for years.

Homes have now been built on portions of the drained wetland. Someone even constructed a storage building over one of the buried drain lines. The area has not been planted for a number of years and still looks dry. Fescue and Johnson grass grow in the unmowed areas, and the ponds are full of water. John tells me that he sees water flowing from the two drain line outlets every time he ventures down by the creek.

It would be difficult to tell that the area was once a wetland. The soil texture on the drained wetland is silt loam and is brown colored. The presence of two small ponds at the base of a hill is a peculiar feature. They are so small that it might be questioned why they were constructed. There are other small ponds at the bases of hills in the area. Perhaps they indicate how others may have copied the successful technique to drain additional wetlands.

I believe the original number and acreage of wetlands present on John D. Smith's land to be similar to many other farms in eastern Kentucky. In driving the countryside around John's farm and surrounding counties, I commonly see signs of drainage just like those on his farm. It is rare to observe or hear of a natural wetland remaining on a mountain farm.

I documented the drainage activities on John's

farm for a number of reasons: John trusted me to describe them accurately; his family had lived on the farm for over 100 years; his recall of land management was excellent; he continued drainage activities his father and grandfather had begun; and his love for wildlife was so great that he encouraged me to record the details so that others could bring similar habitats back to wetlands.

John D. Smith has become a good friend. For most of his life, he operated a backhoe in eastern Kentucky to make a living, which often involved the draining of wetlands for agriculture. John began working with me to restore wetlands at the Daniel Boone National Forest in 1989 and continues to help me as of this printing. With his characteristic dry humor, he said, "The government has been good to me—they paid me for years to drain these swamps, and now they're paying me to bring them back."[6]

Merlin Spencer Farm

Merlin Spencer insisted on driving up the hollow in his four-wheel drive to show his son and me the land his grandfather had farmed in Menifee County, Kentucky. Pointing out a mountainside with a steep 40 percent slope, he said, "When my grandfather bought this farm on August 7, 1897, these hillsides and benches were the only places dry enough to farm. He worked to clear the trees off the slopes and plowed the ground with a pair of mules."[7] Merlin showed me the old plow that his grandfather, his uncle, and his father had used to turn over the mountainsides. He called it a hillside plow. It was made with a latch so that the plow bottom could be turned on a longitudinal axis, allowing soil from the furrow slice to be thrown in the same direction while plowing back and forth across a slope.[8] The hillsides were so steep that a conventional turning plow could not be used, because it was not possible to throw dirt from the furrow slice uphill. As I climbed up the hillside, now grown up to trees, I had to grab branches to maintain my balance. With each steep step, it became clearer why they worked so hard to drain the wetlands that had covered the more level ground along Beaver Creek.

I asked Merlin if he knew what the bottomland fields were like along Beaver Creek in the late 1800s. He said that his grandfather and uncle described it as swampland. There were patches of large trees, areas of water, and grassy places where the top of the water table could be seen in crayfish holes. He said, "You could not walk or ride over these swamps with a horse. It was too wet to farm. In a dry year, you could raise a good crop, but in most years you'd be lucky to pasture an old cow. My dad had to walk alongside the plow because there was too much water for him to go behind it." Merlin explained that if his father plowed the field in the spring, the dirt would form so many clods it couldn't be worked for planting. Therefore, he often plowed in the winter so that freezing conditions would break up the clods in time for spring planting.

It is difficult to believe that the area had been wetland when looking over the now lush, green meadow of grass and clover between Beaver Creek and Highway 1274. The ground was firm and dry underfoot, even though it had been a wet winter and the spring was looking even wetter. I asked Merlin to describe how the 25-acre field with its 1 percent slope had been drained. He explained that his grandfather began the process by removing the clumps of hardwood trees that grew in the swamp. Since a steam-powered sawmill was located just downstream, he didn't have far to haul the logs. He then hand-dug a series of open ditches of various depths and lengths to carry runoff from the mountainside straight into Beaver Creek.

The next drainage action his grandfather took involved even more work. He hand-dug a series of parallel ditches from the upper part of the field down to the creek, and then constructed and buried three long continuous wooden boxes of chestnut in the ditches. One of these wooden boxes was at least 500 feet long. The inside of the channel measured about 6 by 8 inches, with the sawed face of each board facing inward, and the rounded edge of each slab facing outward.

The boxes were nailed together using thick boards of all different lengths and widths. Flat rocks were placed over the top seams of the boxes to keep out soil.

Merlin first hit the top of one of the boxes in 1949 when he was plowing the field with a new tractor his father bought. The tractor plowed at a greater depth than was previously possible with mules and horses. Merlin's son Kim later ran into the top of another section of the wooden drain in 1979 while plowing the same field at a depth of about 16 inches. Both believe that the wooden boxes continue to be functional.[9] The outlet for one of the wooden boxes, where it was exposed to air, rotted and collapsed a number of years ago. They have since repaired it by replacing the section with clay tiles. A few years ago, some of these clay tiles were, in turn, replaced with black plastic drain pipe.

In the early 1900s, the Spencers began using clay tiles to further drain the wetland. Over the years, a number of 3- and 4-inch-diameter round and 4-inch-diameter octagon-shaped clay tile lines have been buried in the field. Seven of these lines were constructed with outlets that empty into a 5-foot-deep ditch that carries water from the mountainside down into Beaver Creek and also separates the Spencer property from the neighboring farm. These clay lines were buried from 3 to 5 feet deep, and they continue to operate.

Merlin explained how he hand-dug ditches to place some of the clay tile lines in the field when he was a boy. Because enough water flowed down the ditches while he was digging them, there was no need to use surveying equipment to maintain grade. The soils were soft, saturated, and collapsed readily, making it necessary to cover the clay tiles as soon as they were placed in the ditch. In quicksand, he lined the bottom of the ditch with flat rocks before placing the clay tiles, to keep them from sinking.

Wes Tuttle attempted to find the site of a buried wooden box in the field with the ground-penetrating radar (GPR) system but was unsuccessful. However, he did find the location of a buried clay tile drainage line. The GPR also re-vealed where a rural water line and a telephone cable had been buried. It is possible that the GPR detected the location of a wooden box, but we did not have enough time to verify what each signal indicated by uncovering every revealed object with shovels.[10]

The drainage actions they completed provided many benefits to the farming operation. They were able to work the fields early in the spring for planting corn. The fields stayed dry later into the fall so that the corn could be harvested. The hay even dried faster on the ground, and there was no longer the concern of having equipment stick in the mud.

His grandfather Jim Spencer was a hauler who transported supplies by horse and wagon most of his life. Merlin recalls that he left home by 2:00 A.M. in order to make it to Olympia 14 miles away in time to meet the 7:00 A.M. train. Jim was always willing to try new ideas and once heard that burrowing crayfish could be killed by dropping salt into their burrows. The high mud chimneys surrounding the crayfish burrows made it difficult to mow a field. The mud chimneys would plug and stop a sickle mower powered by a team of horses. Apparently, Jim tried the salt method and it didn't work. Merlin believes what eventually reduced the number of crayfish from his farm was installing the buried drain lines, which lowered the elevation of the water table over a number of acres.

Merlin explained how his Uncle Gordon removed some of the curves from Beaver Creek to improve the farm. He first used mules and a scoop to start a new channel for the creek along the base of the hillside, and then he took advantage of the force of floodwaters to cut a straight path for the creek. The new stream channel was deeper and carried more water, which greatly reduced the likelihood of Beaver Creek flooding the fields. Merlin indicated where two bends in the creek had been removed, totaling about 600 feet of stream. Shallow, curved depressions still showed where the original creek channel had been located.

Wes Tuttle used the same EM 38 we operated on John D. Smith's farm earlier that day to sample soil conductivity over part of the field that included where Beaver Creek had been moved by Gordon Spencer years ago. The color map Wes prepared on his laptop computer showed a unique pattern that corresponded to where Merlin said the creek had been located. Wes has found that old creek channels will often fill in with fine sediment over time, and that the EM 38 can detect this change in soil texture across a landscape.

The Spencer family has worked for generations to prevent Beaver Creek from moving over and damaging their fields. They regularly remove logs and trees from the creek to prevent its waters from being diverted, and they have worked to armor the banks of the creek with large boulders obtained from nearby maintenance activities along Highway 1274. The Spencers have completed the majority of drainage actions on their farm with no assistance from the government. Merlin thinks that the SCS may have helped them install one drain line in his lifetime of 70 years.

Since 1988, I have driven by the 25-acre field many times without knowing it used to be wetland. While examining the field with Merlin and his sons Kim and Greg, I searched for clues that might betray its existence. One was the huge pin oak with buttressed roots that was growing in the middle of the field. Merlin said that the tree had been there all his life and may have been the only survivor from the original swamp. He was almost thrown from a tractor in 1949 or 1950 when he hit a large root from another water oak that remained buried in the field long after the trunk had been used for lumber.

Perhaps the best signs of the wetland's existence were the exposed drain line outlets in the 5-foot-deep ditch that carried water to Beaver Creek along the lower edge of the field. Kim Spencer had recently cleaned out the drainage ditch with a backhoe and exposed these outlets. They were made of black plastic, octagon-shaped clay tiles, and Orange-Burg pipe (0.5-inch-thick, 4-inch-diameter bituminous pipe used in the 1940s and 1950s). The presence of long, deep, straight ditches along the upper and lower edges of the field also showed how runoff was being di-

Old pin oak shows the location of a forested wetland drained in the late 1800s on Merlin Spencer's farm along Beaver Creek in Menifee County, Kentucky.

verted from the swamp. Soil colors in the field are now mostly brown (10YR 5/3–5/4 Munsell Soil Chart); however, there are patches that contain hydric characteristics (10YR 5/2, 6/2).

John Johnston Farm

John Johnston left Scotland in 1821 to live in America. He eventually bought 320 acres to farm near Geneva on the shores of Seneca Lake in New York. After a few years of farming, he found that the most wheat he could produce was 20 bushels an acre. Water standing on areas of fine-textured clay soils would kill his planted winter wheat by spring.[11]

Johnston had observed tile drainage in Scotland, so he mailed a letter to his grandfather requesting samples. Receiving two clay tile patterns, he asked his friend Benjamin F. Whartenby, a crock maker in Waterloo, New York, for help. Whartenby made Johnston 3,000 clay tiles, and Johnston installed them in a field in front of his house in 1838. This work was done much to the amusement of his neighbors, who would sit on the fence, laughing at what they called "Scotch Johnston's folly." The tiles were placed in the bot-

First area drained with clay tiles by John Johnston in 1838 near Seneca Lake, outside of Geneva, New York. The field is located in front of Johnston's home, across Highway 96A.

tom of hand-dug ditches from 0.5 to 3 feet deep, and then covered with soil. The laughing stopped when neighbors learned that the drained field's yield jumped from 20 to 60 bushels an acre. Johnston was so convinced of the value of tile drainage that he had 72 miles of tile installed on his 320-acre farm by the time of his death in 1880.

Charles Rose Mellen, who purchased the farm

from one of Johnston's daughters after his death, reported that in 1912, an unusually wet season, he successfully planted and harvested a bumper yield of crops from lands drained by Johnston 60 to 75 years before. He boasted about being able to pull a wagon with a 6,750-pound load across a field when, on his neighbors' farms, the soils were too soft for even a mower to cross. Professor John Haswell later reported that the 70 miles of tile installed on the Johnston farm were still operating in 1938.[12]

I visited this historic drained plot in 2003 and found it to be a productive hayfield along the East Lake Road near Seneca Lake. With a gentle 2 percent slope, a farmer could have driven over it without fear of becoming mired in mud or muck. There were no signs of wetlands ever being drained. I believe that this was the first area drained by John Johnston for a number of reasons: (1) The field is shown in the book by Mike Weaver;[13] (2) Merrill Roenke, founder of the Mike Weaver Drain Tile Museum located in John Johnston's home, showed me the field; and (3) Mr. Eddy Kime, whose family farmed the field for generations, confirmed its location for me by telephone.[14]

This first tiled field, or wetland drained by clay tile, is now owned by Lenney Cecere. In 2006, Cecere had owned the field for about 3 years. Mer-

Aerial views of the first areas drained by John Johnston in 1838. (*Left*, color infrared; *right*, black and white)

rill Roenke says that friends had dug up one of the first tiles that a Mr. Wartenby made for John Johnston in this field. The tile is chipped on one end and was given to the Mike Weaver Drain Tile Museum by a Dr. Doran of Geneva, New York, a historian now deceased.[15]

Eddy Kime of Geneva has raised crops on the original Johnston farm for 60 years now, as did his father and grandfather before him. He grows corn, soybeans, and hay on a total of 1,400 acres. Eddy knows that many of the original tile lines laid by John Johnston are still working, including those installed on the first plot of land that Johnston drained in 1838. He says that they did a great job installing the tiles years ago, and he checks the outlets periodically to make sure they are still open. Most of the original tile lines were set at 50-foot spacing, and he has been told that some of the lines were even laid in square patterns to drain especially wet areas. Mr. Kime can still spot some of the old tile outlets along Highway 96A, which goes past Johnston's restored home. These tile outlets empty into the ditch along the road. He looks for green streaks of timothy grass near the outlets, resulting from higher moisture content in the soil.[16]

The soils' textures on the lands being farmed by Kime vary from sand, to loam, to clay. He calls the area a "mix master" to farm. He personally has installed over 100,000 feet (18.9 miles) of drainage tiles on these lands; in addition, his father and grandfather installed other tiles over the years. He claims to be a fanatic about drainage, stating that all of the fields he farms have drain tiles in them.

The soil survey completed for Seneca County in 1972 shows the area first drained by Johnston as

OvA—Ovid Silt Loam, 0 to 3 percent slopes. This soil has a profile that is similar to the one described as typical for the series but is generally a little wetter. Most of the small and medium-sized areas, and a few of the larger areas, occur in positions where they receive moderate runoff from adjacent slopes. Most of the larger areas, however, are on level or nearly level hilltops and receive little or no runoff. These areas have such low gradients that runoff is very slow, and they stay wet for long periods after heavy rains; however there is little or no ponding.[17]

The statement referencing "little or no ponding" should be of interest to wetland restorers. Soil typing of this area was done many years after John Johnston installed the buried clay drain tiles, which he reported, and Kime confirmed, as having a large influence on surface and groundwater. As recorded by Johnston and by what can be observed from color infrared photographs of the area, drainage activities were extensive on and near the original Johnston farm. Therefore, modern soil typing described the changed, non-native characteristics of soils on the site, making it entirely possible that these soils were hydric and subject to ponding before draining by Johnston and the Kime family. Without the buried drainage structures, these soils would become saturated and have water standing on their surface.

Kime claims that, when properly installed, clay tiles will continue to function indefinitely, as long as the outlets are kept open. He tries to check and clean the tile outlets on his farm annually. Clay tile lines he helped install have continued to work for over 50 years. He mentions that water exiting a clay tile line is clear, while water leaving a surface ditch is often filled with sediment.

Asked how to tell if an area has drainage tiles, he gave a simple, concise response. "You'll see a tractor stuck in a field if it doesn't." Sometimes he'll notice slightly healthier and taller crops, or greener streaks of grass growing over tile lines. There are a number of vineyards around Seneca Lake, and Eddy told me that in late 2003, Cecere converted the original field drained by John Johnston into a vineyard. In the fall of 2003, Melissa Yearick, who restores wetlands for the Upper Susquehanna Coalition, visited the area on my behalf. She confirmed that the original drained field had been converted into a planted vineyard and that a winery was being constructed on its center, causing it to appear unlikely that the location would ever be returned to wetland.[18]

Magnified view showing the square pattern of buried drain lines described by Eddy Kime, who now farms the original Johnston farm. The buried drain lines show up as faint white lines on this color infrared photograph taken April 15, 1995, when frost was coming out of the ground.

Close-up view clearly showing a density and pattern of buried drain lines used to eliminate wetlands near the Kimes farm on Seneca Lake.

Linville Adkins Farm

Indian Creek flows down a narrow hollow flanked by steep mountains in Morgan County, Kentucky. Traveling up the winding road cut into the side of a hill, drivers must take extra precautions lest they meet another vehicle. The near landscape—the narrow, gently sloped pasture that lies between the road and the creek—is beautiful. Unknown to most, that view was planned and created by the owner. At one time, the creek meandered over the valley; small streams flowed off the hillside and gradually fanned out, forming wetlands. The natural creek bank had formed levees which pooled waters, and springs provided moisture needed for sedges, bulrushes, and even the occasional willow to survive.

Soon after Linville Adkins bought the farm, he hired Earl J. Osborne to bring in two 1150 Case dozers to change the hollow. Beginning in the dry summer of 1970, the operators cut a deep trench a half-mile long at the base of the mountain and relocated Indian Creek in the straight channel. They moved across to the other side of the hollow and began cutting into the mountain to push huge quantities of soil across the road to cover the wetland and winding creek channel. Earl J. Osborne packed from 4 to 6 feet of soil over the wetlands and placed a gentle slope on the field down to the moved creek. Filling was a slow pro-

(top right) Sloped pasture on the Linville Adkins farm in Morgan County, Kentucky, constructed in 1970 by first moving Indian Creek, then filling shallow water wetlands with 4 to 6 feet of soil.

(middle right) One of the straight ditches that carry water from the mountainside directly into the newly moved Indian Creek.

(bottom right) Barn built on top of the hill from where soil was removed to cover the wetlands once found along Indian Creek.

Pond recently constructed over part of the filled wetland, turned into a field located along Indian Creek.

cess that lasted over a month. As the ground became saturated and softened by the many passes of the dozer, the equipment had to move up and down the creek to let areas dry out. Earl consolidated the intermittent streams that flowed off the mountain into four separate ditches that ran straight to the creek. A flat bench was made on the borrow pit, and slopes were placed on its steeply cut bank.[19]

Earl remembers being paid $10,000 for the dozer work and compliments Adkins on the excellent job he did in sowing the fields to grasses. Adkins even built his barn on top of the borrow site. People who drove by while Earl was working could not believe the changes being made to the hollow. Adkins pastured cattle in the field for years, eventually selling the farm at a considerable profit.

The alterations made to the area were accomplished so skillfully, it is now difficult to believe the hollow was once a wetland. A new house has been built on the upper edge of the field, along with a deep, poorly constructed farm pond in what was once part of the original wetland. The barn blocks the view of where the soil was removed years ago, and dense Virginia pines grow on the excavated slopes.

While visiting the site with Earl, I pointed out the presence of a metal pipe next to the moved creek, which served as an outlet for drain lines buried in the field. Earl had no knowledge that a buried drainage system had been installed in the filled wetland. The presence of the buried drainage system and the fact that Earl had trouble with soft ground shows that the wetlands had also been fed by springs. The springs eventually saturated the soils Earl placed over the site, making it necessary to remove excess waters with an underground drainage system to enable farming.

Primary signs that indicate the area was modified are the presence of ditches that run perpendicular between the hill and Indian Creek, along with Indian Creek itself. Indian Creek is now straight and flows along the base of the mountain for over 0.5 mile. The metal pipe emptying water into the moved creek also shows that the location was once very wet.

Corbin Wetland

Sometimes unexpected attention to a wetland can lead to its demise. In 1993, a team of botanists, biologists, and zoologists set out to inventory rare plants and animals in the Stearns Ranger District of the Daniel Boone National Forest. While driving Highway 25W in Whitley County, Kentucky, team members spied a natural wetland on private land. Hoping to discover the rare white fringeless orchid on the site, they stopped to investigate. Even though the rare orchid was not found, they informed the landowner of the potentially unique importance of the wet meadow wetland. Apparently, the landowner did not share their excitement, as a short time later, a dozer was used to construct a go-cart track, open water pond, and topsoil pile on the wetland, thereby obliterating the site. The experts involved could not help but believe that the unexpected interest they gave to the wetland prompted the landowner to remove what appeared to him as a possible liability on the property.[20]

Marsh Branch

The steep hills of eastern Kentucky contain some of the smallest and most unusual wetlands on the Cumberland Plateau. A low-volume perennial stream that flows down a mountain ridge in Jackson County on the Daniel Boone National Forest forms one of these unique wetlands. With no defined channel, the stream consists of a braided ribbon of bulrushes, sedges, and sphagnum mosses growing on soils kept saturated by a subsurface flow of water. Occasional small pools of water, formed by logs interrupting the flow of groundwater, provide miniature breeding sites for mountain chorus and wood frogs. Scattered red maple, white oak, and shortleaf pine grow near the wetland, and alder shrubs dominate the canopy.

This wet meadow wetland that occurs at the source of the stream provides critical habitat to what may be the largest remaining population of the white fringeless orchid in Kentucky. Unfortunately, the wetland was probably changed by construction of both a transmission line and a distribution line over the site in the late 1900s. The actual effect of cutting trees for the path of the power line (right-of-way) on the wetland is not known. However, continued vehicle access and maintenance threaten both the rare orchid and the wetland itself.

About 4 years ago, a channel began forming at the low end of the wetland, apparently the result of trucks and tractors attempting to drive around the saturated soils. The new channel concentrates flowing water, increasing its velocity, leading to erosion that cuts farther uphill with each rain. The ground is firm next to the channel, proving it is working as a ditch to remove water from the surrounding soils. Typically, the erosion caused by the flowing water would be halted by large trees falling over and forming small dams across the channel to block it. With only saplings left growing in the right-of-way, large trees are absent.

Unfortunately, orchid numbers drop as the channel grows in length and depth. Climbing

Pond, go-cart track, and topsoil pile on a natural stream-head wetland constructed by a private landowner soon after a team of resource managers expressed interest in the site near Corbin, Kentucky.

fern and Asian mint, both invasive species that favor drier soils, now advance across the wetland, crowding out even more orchids. There has been conflict regarding how the wetland should be managed. The East Kentucky Power Cooperative firmly declares that there is a strong need to prevent trees from growing in the right-of-way in order to maintain reliable electric service. The piles of dirt placed across an access road by the USDA Forest Service to block its use are being skirted by vehicles, and debate among managers on how best to restore its hydrology continues as the wetland dries.

Centers Hollow

It is obvious that Donnie and Betty Centers have worked hard to build their hobby farm in the mountains of eastern Kentucky. The narrow, grown-up hollow they bought in 1990 is now neatly mowed and provides enough room for the home and four buildings that Donnie built, one of which serves as Betty's tax office. Donnie retired in 2004 from a career with the Forest Service

Spring

(top) Raked hay marking the center of a destroyed wetland on the Donnie and Betty Centers farm in Morgan County, Kentucky. Cattails grow in the ditch dug for changing the creek. Large amounts of soil were moved from the mowed hillside to cover the wetland.

(bottom) Spring that supplied water to drained wetland betrayed by a puddle of water on the Donnie and Betty Centers farm. The spring emerges from under 4 feet of fill to escape capture by the buried drain lines.

me that the hayfield was a "low place in the bottom that had pools of standing water in it" when he bought the hollow. Willow and red maple trees grew around the edge, with cattails and bulrushes flourishing in pools of water. Before his drainage actions, Donnie could not farm the spot, and he doubts that anyone else ever had.[21]

In 1991, Donnie hired Billy Osborne to use his TD15-C International dozer to reshape the hollow to meet his needs. Billy leveled a flat area at the foot of the hill for Donnie to build his home and straightened the meandering creek to provide a larger lawn. Donnie intended to build his driveway through the center of the hollow, directly over the wetland. Billy suggested that they instead place it at the foot of the hill, so it wouldn't be as likely to sink.

To eliminate the wetland, Billy first excavated a deep ditch along the base of the hill and redirected the creek over into the straightened channel. He moved a considerable volume of soil from the base of the mountain over the future driveway's location and into the wetland. He eventually pushed from 4 to 5 feet of soil on top of the standing water. When grading the final road, he cut a ditch along the base of the hill to divert water from the lower ground.

For years, Billy believed that his dozer work had been sufficient to dry the location, and he was surprised when Donnie told us that the area remained so wet that later he had hired a backhoe and operator to bury drain lines in the new field. They placed 4-inch-diameter corrugated plastic drain lines in ditches 2.5 feet deep that started at the base of the hill and emptied into the deepest section of the moved creek near the main road.

As we looked over the small hayfield, I asked Donnie and Billy if they could point out any signs to me that would indicate that the site was once a wetland. After a bit of thinking, Billy mentioned "the straight creek at the foot of the hill" that emptied into the deep ditch along the county road, and the "cattails in the ditch," which they both felt were left behind from the original wetland. They also pointed out the steep slope above

and then devoted most of his time to farming. He described the buildings that he constructed with the same pride he displayed when we walked over to a hayfield to discuss its transformation from what he called "the swampy place." Donnie told

the wetland where it could be observed that the toe of the mountain had been cut off to supply soil for filling the wetland, with a long bench being made along the base of the mountain for the driveway. I also noticed a small willow and a red maple, unusual trees for the immediate area, that were growing behind Betty's tax office as further signs the area was once wetter.

As I walked over the freshly cut hay, I found a saturated spot about 6 feet in diameter in the middle of the field. Donnie said that the wet place was a spring that appeared for the first time in May 2005, following record amounts of rainfall. Apparently, the hydrology of the wetland had been maintained by a combination of hillside runoff, an intermittent stream, and the spring. Even though Donnie had diverted the hillside runoff and moved the stream, the spring continued to flow, eventually saturating the soil in the wetland. His buried drain lines had carried excess water from the soil for the past 13 years, until this spring. Donnie did not appear very concerned about the spring's emergence and said, "It won't be much trouble to fix it with another drain line."

Carter County Oxbow

Greg Maddix's home overlooks what was once a 3-acre oxbow wetland on the Little Fork of the Sandy River in Carter County, Kentucky. In 1998, Maddix approached Jimmy Lyons, district conservationist for the NRCS, for assistance in draining the emergent wetland. Maddix described it as "nothing but a mosquitoes' nest, so he wanted to get rid of the water so the kids didn't get West Nile Disease." With no cost-share money available, Jimmy designed a simple drainage system that Maddix proceeded to install with his own money. He hired a backhoe operator to dig a ditch from the deepest portion of the oxbow down slope to the river. A 6-inch slotted plastic drain pipe was buried in the ditch to carry water from the wetland to the deep outlet, which was the river. He

(top) Orange Hickenbottom structure provides clear evidence that a substantial amount of water is being removed from this drained wetland on Greg Maddix's farm in Carter County, Kentucky.

(bottom) Rushes and sedges are present in what was once the deepest part of the historic oxbow wetland on Maddix's farm.

then attached a Hickenbottom surface inlet drain structure to the drainpipe and placed a load of #2 limestone gravel around the base of the structure to prevent the holes from plugging with soil or vegetation. Jimmy said that the rock would also maintain waterflow should the upright ever be

Complex of wet meadow, scrub-shrub, and ephemeral wetlands drained by a series of excavator-dug ditches in order to raise cabbage outside the community of 100 Mile House in British Columbia. The ridge of soil that parallels each ditch, along with the pools of standing water, indicates an incomplete and poor-quality drainage job.

owners who may never know what the original oxbow looked like. An orange Hickenbottom provides evidence that the site once contained an abundance of surface water. The area would make an easy restoration project, as all that would have to be done is to use a backhoe to find the buried drain line and then block it as it leaves the historic wetland. The shallow overflow ditch that formed between the oxbow and the river after the outlet was excavated would also need to be filled in, restoring the natural levee along the river, to allow the water depth to return to its original level. I pondered, as I drove home, that even should a restoration ever be attempted, agencies might balk at approving the permit application in a misguided, yet sincere, attempt to protect that small portion of the area that still qualifies as a wetland, even though it is in a greatly changed condition from its original state.

The Cabbage Patch

Scott Whitecross drives past the old cabbage patch every day on his way into the town of 100 Mile House, British Columbia. The view over this large expanse of fields surrounded by mountains where produce was grown is nothing short of spectacular. Scott became aware that the fields he had been looking at for years were once hectares of wetlands while attending the Wetland Institute sponsored by the British Columbia Wildlife Federation.[23]

Scott remembers when a family from Europe purchased the large ranch in the early 1990s. They had been told by the previous owners that the fields never flooded, only to find that in the wet spring of 1996, much of the area become a seasonal wetland. Scott recalls seeing an excavator working to dig open drainage ditches on the ranch in the winter and early spring. People in the community noticed the ditching going on that spring, mostly with indifference, as the area was not considered to be of much importance. The ditches dried out the wettest part of the wetlands, allowing for the cutting of hay in dry times.

broken. The backhoe did not fill the oxbow, so the wetland's original shape remained basically unchanged. Jimmy stated the drain made quite a difference. The area now holds only small pools of water when high flow in the river prevents water from leaving the wetland. The wetland is now dry enough so that two-thirds of it can be cut for hay. Before the drainage, the owners had trouble accessing a field on higher ground along the river; however, this has not been a problem since the project was completed. "The oxbow could have been dried up enough to farm the whole area with some contouring and a couple more drain lines," said Jimmy.[22]

Bulrushes and sedges now grow in what was once the deepest part of the wetland up to and around the Hickenbottom structure. Water could be heard flowing down into the structure 2 days after a heavy rain. Hydric soils could be found where aquatic plants were growing in what was now the deepest part of the depression. A recent aerial photograph clearly showed the original curved shape of the now dry oxbow wetland.

Maddix sold his house in mid-2005 to new

Straight, open ditches in a grid-like pattern combined with black, organic soils betray the fact that the area was once a wetland.

Scott's older neighbors say that most of the cabbage patch was willow flats with sedge and grass meadows that were seasonally flooded by the creek in the valley center and from high waters from the adjacent lake prior to 1900. Beginning in 1920 and continuing into the 1960s, the willows were gradually cleared by burning, the use of horses, and later by tractors and bulldozers. This allowed hay cutting on the greater portion of the meadow in most years; however, in some years, large areas would remain too wet to cut. The new owners have since moved back to Europe, yet the network of ditches they made continue to drain the wetlands.

Vestiges of
Wetlands Long Past

Identifying drained wetlands can be difficult. When a thorough job is done in draining, there is sometimes no indication that an area was once a wetland. In some cases, any traces left are barely noticeable. Don Hurst, district conservationist with the USDA Natural Resources Conservation Service (NRCS) who has devoted a career to helping farmers drain lands as a means of improving crop production, says that he would have a tough time spotting an area that used to be wet, providing a quality job was done with the drainage, as few, if any, signs remain to betray the previous nature of the site.[1]

Basically, a destroyed wetland no longer looks like a wetland. Gone are the aquatic plants, standing water, groundwater, and revealing gray-colored soils. My interviews with those who have drained wetlands indicate that, once started, they usually finished the job. Drained wetlands now look like any other agricultural field, woodlot, or housing development.

Most wetlands were drained so many years ago that there is no written or verbal record of their presence. The SCS estimated that, by 1950, over 50 million acres of wet farmland had been drained across the United States by individuals in private projects not associated with public drainage enterprises.[2] In many areas, the few natural wetlands that remain are those that were too deep to drain and too large to fill. In working to identify wetland restoration sites, most individuals look for basins that are still growing aquatic plants, with saturated soils, and an obvious ditch running down the middle. Such areas generally represent partially drained and failed attempts to convert a wetland.

A number of drained wetlands were quite small, less than 1 acre in size. This is logical, considering that early agricultural fields in the mountains were also diminutive, and there was much to be gained from farming a level piece of ground. In his book about land drainage, John Klippart provides data on clay tile supply needs and distances between drains for sites as small as 0.25 acre.[3] Many of these smaller drained tracts are no longer used for farming and have since grown up to trees, further masking signs of historic wetland presence.

Fortunately, there are clues that betray the presence of drained wetlands. I have summarized these signs after interviewing individuals who worked to drain wetlands all their lives and by examining known drained wetlands in a number of states and provinces.

Written Drainage Records

It is rare to find written records of drainage practices on a farm. Those who did this work were often too busy to record such activities or may not have had the skills and survey equipment needed to develop accurate maps. The best opportunity to find any documentation of drainage activity is from those projects that were funded by the government. The Natural Resources Conservation Service, formerly the Soil Conservation Service, actively designed and funded drainage activities for most of its existence. Detailed plans were often prepared for drainage projects and can be found on file in NRCS offices. However, I'm told that files are updated and cleaned after a number of years, so it may be difficult to locate these old records as some may be archived in central offices. This is unfortunate, as I have seen sites where the SCS designed and funded some of the most effective drainage practices ever completed in North America.

Small Unplanted Areas in Fields

One clue to the presence of a drained wetland is seeing an area in a field that is not planted or mowed every year. Trees and shrubs growing in these locations often represent the deepest portions of much larger wetlands drained years ago. Rocks, logs, and brush may be piled in these areas as they are considered to be nonproductive.

Springs

Finding a spring that does not saturate the surrounding ground, or whose overflow does not form a channel leading to a creek, may indicate a drained wetland. Springs are responsible for maintaining many wetlands; farmers recognized this fact and took measures to carry runoff from springs beneath the ground directly to streams, thereby draining these wetlands.

Verbal History

Talk to residents who really know the land, especially those who raised crops in an area for years. Ask if they know of places that used to have puddles of water or that stayed wet into late summer. Inquire about the location of swampy places, as the current definition of wetlands is unclear to many. A place where they may have gotten their tractor stuck often indicates a drained wetland.

Brown area of exposed soil where planted corn drowned in the extraordinarily wet summer of 2003 marks the location of a wetland eliminated with buried drain lines in Trigg County, Kentucky.

Dry Level Ground

Undisturbed level areas with fine-textured soils should show wetland characteristics such as pools, hydric soils, and aquatic plants. Locating a dry area that is level or gently sloped without wetland characteristics that occurs on silt loam, silt clay loam, or clay soils often indicates a drained or filled wetland. The lack of wet depressions in a field or abandoned field provides strong indication that land leveling and filling activities took place to eliminate wetlands.

Long Deep Ditches

The presence of long deep ditches indicates that wetlands have been drained over a large area. These ditches are basically straightened streams designed to carry large amounts of water. They often cross lands owned by a number of property owners and may now appear quite natural, being bordered by large trees and containing aquatic vegetation. Some are even labeled as streams

(top) *Before:* A corn belt pothole lake as it looked in early spring. (bottom) *After:* The same spot, in August, drained, growing corn 9 feet high. (From R. A. Hayne, *Drain the Wet Land* [Chicago: International Harvester, 1921], 45)

on topographic maps. These deep ditches were generally not the cause of wetland drainage but served as the means by which a large number of individual wetlands in the surrounding landscape could be drained. The ditches provided the outlet essential for buried drain lines to remove water from depressions and level areas. These ditches were often the first and most important action that enabled landowners to remove water from their wetlands without fear of flooding their neighbors downstream. In the mountains, these ditches may be less than 0.5 mile long; on more level terrain, they can run for miles.

On a rainy December morning, Richard Bond Jr. guided me to Hannah Lane in Carter County, Kentucky, to see a field he had drained with clay tiles back in 1965. Looking out over the lush green of the field, he said, "That place was a swamp. You could only farm a couple of high spots because it was all wet." Richard changed all that. The SCS had sketched a drainage plan for the field, owned by Homer Womack and Sons, for Richard

to follow. With the help of two laborers, Richard used his backhoe to install thousands of 5-inch-diameter clay drain tiles in a grid-like pattern over the field. Tiles were placed in ditches from 30 to 60 inches deep, with only one outlet being dug. The outlet ran water into a deep community ditch that had been excavated with a dragline years before. Even with the record amount of rain we had been having, the field was dry and looked as though it could be driven across without fear of leaving the vehicle stranded halfway.[4]

I asked Richard how in the world you could tell that the area had been drained. He said, "Look how green the grass is, and then look across the road." On the other side of the road there was another field, its grass a bit yellow, and water stood in many of the shallow open ditches that covered the field. He said that originally both of these fields had looked the same; however, "the farmer on the other side didn't use tiles to drain his swamp, and you can tell." Henry French tells a similar story of traveling through Wales with

English farmers when the slightest variation in the color of wheat and corn would attract their immediate attention. Noting yellow or light-colored crops, they would remark that the field was in need of further drainage.[5]

Richard then pointed out how important the deep open community drainage ditch had been to the project: "That gave us the drop for the outlet. Otherwise, there would have been no place to send the water." Before we left, I kidded him about a puddle we could see near the upper edge of the field. Showing genuine concern, he said, "That water was more than 100 feet from a drain line, but we could have got it with an extension."

Shallow Ditches

Look for the presence of ditches, even shallow ones. Ditches carry surface water that once formed wetlands directly into streams and rivers; they also lower the water table. A field or woodlot containing a series of parallel ditches from 50 to 100 feet apart was probably a wetland. Parallel ditches can be difficult to see after an area has grown up to shrubs and trees. Ditches only need to be a few inches deep to divert water, and some carry water only after a heavy rain. They were often dug along the base of a hill to keep runoff from maintaining wetlands on more level ground. Many were designed to be 30 feet wide or more with gradual side slopes so they could be crossed by heavy equipment. Ditches were also used to divert water from wet meadow and shrub wetlands found on sloped lands.

Finding a series of shallow ditches that parallel each other at only 30- to 60-foot intervals shows that an area was once so wet that directional plowing was used to create ridges for planting crops. The peak of the ridges may be 1 to 2 feet higher than the bottom of the ditches, and these ditches may or may not empty into a deeper ditch or creek. This practice of plowing, known as "lands," was used to raise crops on areas of wetland and generally indicates that more effective buried drain lines are not present. Narrow lines of

Deep open ditch, constructed with a dragline, provided the drop needed for draining many acres of wetlands near Hannah Lane in Carter County, Kentucky.

bulrushes and sedges will often grow in the shallow ditches between the slightly higher ridges of soil. These "green streaks" of aquatic plants may parallel each other over the entire length of a field, often disappearing when the field is plowed or cut for hay.

The lands pattern is more difficult to detect when an area is no longer farmed. Trees and shrubs soon grow in the bottom of the ditches, and they may look like overgrown fence rows. These rows of trees may even be present in the ditches when the area is being farmed, generally indicating that an individual with less concern for the productivity of the field is now tending the farm. Keeping trees from growing in the bottom of the ditches along the base of lands requires more work than cutting hay or harvesting corn on a tractor. The farmer has to pay close attention to the weather and mow these ditches when it's dry enough not to get his equipment stuck. In wet years it is necessary to cut the trees by hand, for if trees are allowed to grow in the ditches, they will soon reduce crop production by shading the field.

Shallow ditches were often used to remove surface waters in order to provide a seedbed for growing more commercially valuable pines

Remains of a drained wetland along the Licking River in Morgan County, Kentucky. A deep ditch, now grown up to trees, drains the wetland by cutting through the natural levee along the river. The ditch was only a small part of the original wetland.

The presence of numerous shallow ditches in this field in Bath County, Kentucky, is a strong indication that the area was once a wetland.

Open, farmed ditch removes a large amount of water from a hayfield in Bath County, Kentucky. Note the area too wet to mow on the right side of the ditch, downhill from the first two wrapped bales, which may be caused by a plugged underground drain line.

These small ditches continue to remove runoff from a wetland originally drained for agriculture, now grown up to trees in Trigg County, Kentucky. Notice that the ditches join at right angles and can be readily seen after the hard rain.

Now partially filled by sediment, this open ditch was dug to drain surface water from a forested wetland near the Red River in Powell County, Kentucky.

throughout the South. Most of the ditches used to drain areas for pine were not maintained and are now hidden by trees and shrubs, yet they continue to function. The shallow ditches used to drain wetlands for pine can be spaced as much as 0.5 mile apart and often empty into constructed ditches at least 4 feet deep. Roads were often built along the edges of these deeper ditches from the soils wasted from creating the ditch. Culverts were positioned beneath the constructed roads to transfer waters from the shallow ditches on the other side into the deeper channel.

Pines are also planted in wetlands by directionally plowing saturated soils to create lands. This practice, often called "bedding" in forestry, provided a series of long narrow ridges whose tops rise above surrounding water. Pines planted on these ridges will survive in spite of standing water and, as they grow, will dry the surrounding soils.

Finding an open ditch that ends before reaching a creek often indicates the presence of a drain line buried in the ditch. The buried drain line carries water from the ditch and drained wetland downhill to a creek. These short sections of drain lines were often used as culverts to provide equipment access to fields.

Definitely the presence of open ditches indicates the historic presence of wetlands. In 1903, Henry French made the following comment about using ditches: "Open drains are thus essential auxiliaries to the best plans of thorough drainage; and, whatever opinion may be entertained of their economy, many farmers are so situated that they feel obligated to resort to them for the present, or abandon all idea of draining their wet lands."[6]

Gaps along the Riverbank

I find it worthwhile to walk along a creek or riverbank to examine how side drainages enter the main channel in order to identify possible drained wetlands. If a lower-order stream enters the main channel at a right angle, and its channel is rather straight, it most likely represents a constructed outlet for draining a larger acreage. These outlets were designed to drain standing water trapped behind the natural rise in the creek bank. In comparison, the unaltered creek's entrance is wider and winding, with more gradually sloped banks.

Doreen Miller has been working to establish bog habitats for over 15 years in the Nantahala National Forest in the mountains of North Carolina. As a wildlife biologist, she is charged with increasing populations of the endangered bog turtle and other rare species that depend on these imperiled ecosystems. Doreen has spent years controlling trees and shrubs that shade sphagnum mosses, cutting and spot burning the woody vegetation that lowers a water table and dries out mud important to the turtles. One of the largest remaining bogs on the Nantahala is found along the banks of the Nantahala River. When I visited the site, beaver had recently dammed a small stream that bisected the bog, improving the hydrology of the site. Unfortunately, large portions of the bog remained dry and dominated by woody vegetation.

On a beautiful spring day I planned to examine parts of the 7-acre bog needing attention with Doreen, only to find that we had to crawl beneath a dense tangle of shrubs to make progress. Instead of fighting the brush, we decided to walk the river bank bordering the lower edge of the bog and discovered two factors responsible for its continued dryness. The first was a small stream passing through a campground along the upstream edge of the bog. The stream had been channeled and moved directly into the Nantahala River so that it no longer entered the bog. The second was a 45-foot-wide ditch that cut through the natural rise or levee along the riverbank. The ditch acted as an emergency spillway on a pond, controlling water levels over the entire bog. Apparently, efforts had been taken years ago to dry out the lands along the river so they could be farmed. These early drainage actions continued to remain effective 60 years after the lands became national forest. Doreen had walked the riverbank along the edge of the bog many times and even remembered get-

Doreen Miller in a constructed ditch that drains water from a bog directly into the Nantahala River in North Carolina. The ditch was constructed to help drain the wetland years before the land became part of the Nantahala National Forest.

ting her feet wet as she crossed the shallow water flowing from the flat-bottomed ditch. The wide ditch blended in with the surroundings so well that it appeared to her and many others to be a natural feature.[7]

Considerable success has been realized for hundreds of years in the conversion of bogs to farmland. In the 1919 book *Engineering for Land Drainage,* Charles Elliott devotes an entire chapter to the drainage of "Peat and Muck" lands. Techniques involving the use of open ditches and buried clay tile systems are outlined for draining bogs situated on clay or sandy soils. One point of interest was that drainage systems were installed in order to maintain the water table approximately 24 inches below the surface, which was deep enough for cultivation yet high enough to avert significant settling.[8]

Straight Streams

Streams that look basically straight with few curves and bends were probably made that way by people working to create large, unbroken

fields. Natural meandering streams and wetlands go together, unfortunately, as do straightened streams and dried lands. Expect natural wetlands to have been eliminated from an area if streams have been channeled or straightened.

The consequence of a stream being straightened is generally the lowering of the water table on adjacent lands, which can act to drain riparian wetlands maintained by an elevated water table. A straightened stream functions the same as a deeply cut ditch to lower groundwater and dry soils for planting. John Johnstone in 1808 described how groundwater-supported wetlands originally formed by a river changing its course could be drained by deepening and widening the adjacent bed of the river.[9]

A standard method for draining wetlands in mountainous areas involved moving small streams that flowed off hillsides into open ditches that had been constructed to traverse the shortest distance from the base of the hill to the main creek. Basically, runoff from each hollow was directed into its own straight ditch, which entered the main creek at a right angle. Many wetlands that were once supplied with water from these small streams meandering over bottomlands were drained by this technique. Sometimes the dry basins that showed where these wetlands were located remain visible; however, in most cases, they have been leveled and filled.

In an 1891 book on drainage, A. I. Root (publisher and author of the appendix) went into great detail describing how he fought to straighten Champion Creek in Ohio, which resulted in changing low wet ground into a market garden. Using horses, plow, and scraper, he cut a straight channel for the creek, which in one place involved cutting through a bank that was 6 to 8 feet deep. Upon completing the project, he realized that in the future he could save work by simply digging a straight shallow ditch between bends and then waiting for the stream to complete the job. He identified many advantages to his battle with the stream, which included producing better-shaped fields for farming, reducing crop damage from flooding, and stopping the creek from creating new channels. In summing up the story he asks, "Did it pay? Well, you should come down to my creek-bottom ground when we are gathering crops, and see the wonderful growth that we have right in these very low places, where the frogs and toads ten years ago used to hold 'high carnival' during the greater part of the season."[10]

The practice of straightening streams so they could serve as channels for improved drainage was recommended by Elliott in his 1919 book about land drainage:

A crooked channel may be greatly increased in carrying capacity by cutting across the bends in such a way that the water will flow in a fairly straight line down the valley, provided the size of the channel throughout is properly adjusted to the new condition. . . . This method of improvement should be consistent throughout the valley, otherwise the relief of one part of the stream may result in the congestion of the water and consequent overflow of lands in another.[11]

A drawing shows that it is often necessary to straighten a stream over a considerable distance to avoid overflow problems if only a few bends are removed from the creek.

French recommended straightening and clearing natural streams in order to create outlets that would be deep enough to facilitate the installation of buried lines to convert swamps. Two agricultural engineers mailed a letter to French describing a drainage project they completed near Boston in 1859: "An outfall was obtained, at the expense of considerable labor, by deepening the Roxbury and Dorchester Brook for a distance of nearly a quarter mile, about four hundred feet of which was through a rocky bottom, which required some blasting."[12]

Small streams were often directed underground as a way to improve fields. Root described how he placed a "wet weather stream" that crossed his land into a series of buried 12-inch clay tiles in order to protect the strawberries and raspberries he was raising in Medina County, Ohio.[13]

The moist depression and meandering basin, once wetland, were drained by moving the small creek into a ditch along Highway 205 in Morgan County, Kentucky.

Aboveground Pipes

A vertical, short section of pipe with holes in it above the ground generally marks the presence of a drained wetland. Known as a Hickenbottom surface inlet, these are used to rapidly drain surface water from a wetland.

Field created by draining wetlands with a series of open waterway ditches, each with plastic 4-inch drain lines buried in them connected to Hickenbottom surface inlets. A Hickenbottom can be seen in the background of this photograph taken in Wolfe County, Kentucky, along the Red River.

Buried Drain Lines

Finding clay tiles or plastic pipe is a strong indication that the area was a wetland of some type. Drain lines are buried 30 inches or more in the ground to remove both surface and subsurface water from the soil. They often start uphill from a wetland, pass through the wetland, and carry water downhill to an outlet, which can be an open ditch, stream, or river. Drain lines effectively lower the water table, remove standing water from large and small areas, and can remain effective for 100 or more years with little or no maintenance.

Since he had over 60 years of experience in designing and installing drainage systems on farms in western Kentucky, I asked James "Booster" Flowers to give me some tips on how one might determine if an area had been drained with underground clay tiles or plastic pipe. He said to "look for the first place to dry out after wet weather because it's closest to the buried drain line. You might be able to find clay tiles with a steel probe, but you'll never find plastic lines that way." He went on to explain that it was common to find buried clay drain tiles from previous installations in wet places when installing new drainage systems for the SCS.[14]

Single drain outlet in an 8-acre field with over 6,000 feet of buried drain lines in Powell County, Kentucky, on the Seldon Reed farm. Outlets like these can be difficult to locate.

Sometimes you may get lucky and actually find a drain tile outlet by walking the creek bank. I've attempted this with only limited success. Creek banks are usually overgrown with shrubs and so steep and muddy that they are difficult to thoroughly examine. An outlet may be visible as an exposed pipe with water flowing from it or be simply a wet spot 3 feet or more below the top of the creek bank where it was buried. Finding a tile outlet indicates that extensive measures were taken to drain lands uphill from the site. Large drainage systems are typically emptied into a single outlet in order to reduce maintenance costs and prolong the life of the system. Therefore, locating an outlet proves that wetlands have been drained upstream from the site.[15]

Buried drain lines will continue working even after an area has grown up to trees. Henry French wrote: "The behavior of roots is, however, very capricious in this matter; for, while occasional instances occur of drains being obstructed by them, it is a very common thing for drains to operate perfectly for indefinite periods, where they run through forests and orchards for long distances."[16]

Buried drain lines can be difficult to detect without extensive excavation. However, a number of experiments have been conducted to locate buried drain lines by using ground radar with limited success. The radar units, which are calibrated to detect possible underground channels, are attached to trailers pulled over the ground,. This research has been focused on helping farmers locate buried lines in need of repair, and it may also help individuals find any buried lines that are responsible for wetland failure.

Fluorescent Flags

Farmers often mark the surface inlets in their fields so that they can avoid damaging them while planting and harvesting. Generally using fluorescent orange flags atop long poles, these markers are readily seen in the bottom of large and small depressions that were previously wetlands. Water enters the surface inlet near the marker and is then carried beneath the ground via a buried drainage system to an open ditch or creek some distance away. The flags used to mark surface inlets can be spotted on both high and low ground.

Wells

An open or covered well may indicate the presence of a drainage system used to eliminate wetlands. Deep wells were dug to serve as sediment traps for tile systems and as places to combine waters flowing from smaller lines into a larger line. Wells can be of various sizes and may even have been constructed large enough for a person to enter for cleaning. These old wells or drainage sumps have been found on National Forest System lands now grown up to trees, and until further research was done, it had been incorrectly assumed that they had been dug to provide drinking water at historic home locations.

Aerial Photographs

Aerial photographs can help show where drained wetlands occurred on a landscape. The straight lines of open ditches are often visible on aerial photographs, especially when an area has not grown up to trees. Observing a series of parallel straight lines in a small area can indicate where ditches were used to drain a wetland. The practice of lands used to farm wetlands by plowing to create a series of parallel ridges can be seen on old aerial photographs. An infrared aerial photograph may show the presence of buried drain lines in northern areas if taken when winter frost is coming out of the ground. However, once the land is plowed or growing vegetation, these buried drain tiles will not be visible.[17]

Landscape patterns can provide clues to the historic presence of wetlands in an area. Aerial photographs clearly show the outline of square-mile units of land, known as "sections," bordered by roads in the agricultural region of the Midwest. It is not difficult to find sections that contain wet-

Orange flags at two surface inlets used to drain a wetland along Highway 7 in Chippewa County, Minnesota. The open ditch *(left)* provides an outlet for the buried drain lines that begin at the surface inlet *(right)*. The soybeans' yellow color indicates stress caused by excessive moisture in the field.

lands adjacent to those that do not. Those sections that contain wetlands can easily be seen from an aircraft at an elevation of 35,000 feet when flying over the prairie pothole region in South Dakota and Minnesota. Sections that lack wetlands often indicate farms where landowners have taken action to remove them. By looking closely at those sections that do not have wetlands, the darker-colored imprint of eliminated wetlands can often be seen in agricultural fields.

Comparing recent aerial photographs with photographs taken years ago can help identify drained wetlands. The presence of shallow water and noncultivated areas surrounded by fields may indicate wetlands on old black-and-white photographs. In some areas, it is possible to find aerial photographs taken in the 1930s. These old photographs may be stored at USDA Natural Resources Conservation Service, USDA Forest Service, and other government offices.

Dry Depressions

Observing depressions of fine-textured soils that do not contain surface water or which lack hydric soils often indicate drained wetlands. Dry basins that were once wetlands can range in size from less than 0.25 acre to as much as 300 acres. The surface and soils in these low areas are generally kept dry by buried drain lines or even a vertical

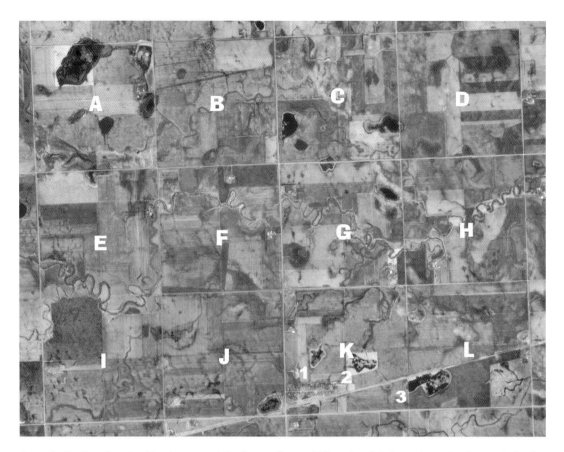

Strong indication of wetland loss in an area 0.6 miles northeast of Albee, South Dakota, shown by changes in landscape pattern. Sections marked A, C, and K contain a number of wetlands, while the majority of wetlands have been drained in Sections B, D, E, F, H, and J by the use of surface drainage ditches. Existing wetlands are labeled 1–3 on Section K.

well, which, in turn, indicates a drained wetland.

Drained bogs can also appear as dry depressions on the landscape. A number of authors discuss how much settling can occur when a bog is drained. Bogs were often dried with underground drains made of clay tile, wood, or stone.[18]

Finding a bottomland area of dry, silt-loam-texture soil where a natural levee parallels the stream often indicates a drained wetland. Natural levees can act like dams, often forming wetlands between the stream and hillside. Landowners buried drain lines in these areas so that they passed under the natural levee to transport water directly to the stream, thus draining wetlands.

Once an oxbow wetland, an abandoned channel of the Red River was drained in 1985 by burying a 6-inch-diameter corrugated plastic pipe with a laser-guided tiling machine in Powell County, Kentucky.

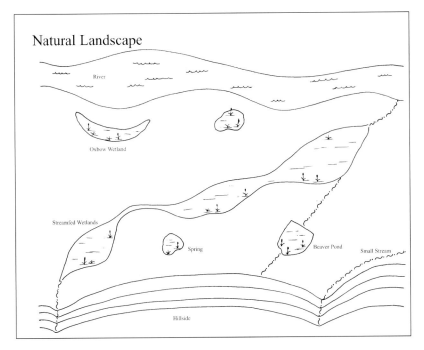

Natural Landscape

River

Oxbow Wetland

Streamfed Wetlands

Spring

Beaver Pond

Small Stream

Hillside

112a, b. Comparison of a natural *(top)* with a drained *(bottom)* mountainous landscape. (Drawings by Dee Biebighauser)

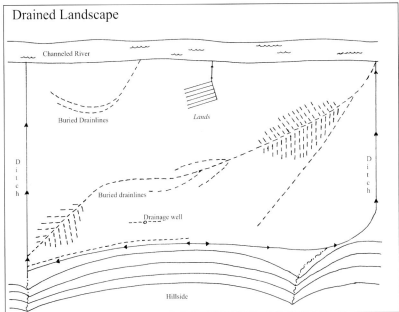

Drained Landscape

Channeled River

Buried Drainlines

Lands

Ditch

Ditch

Buried drainlines

Drainage well

Hillside

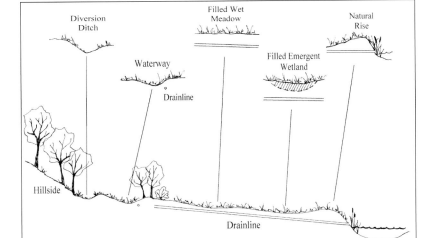

Diversion Ditch

Filled Wet Meadow

Natural Rise

Waterway

Filled Emergent Wetland

Drainline

Hillside

Drainline

Profile view of a drained landscape. (Drawing by Dee Biebighauser)

95

Small Wetlands

The presence of small wetlands often shows where much larger and different types of wetlands were located. Small wetlands often form in the low places of open drainage ditches and road ditches that were used to dry larger systems. Small, saturated patches of ground often indicate the location of a single blocked drain line that was used to eliminate a much larger wetland. The pressure in a drain line can build uphill from the blockage and force water to the surface through gaps between the pipes. French described a method for finding obstructions in buried drain lines by punching a hole in the soil along the course of the drain. Water will burst up from the hole like a spring at the point of greatest pressure just uphill from the blockage, while there will be no upward flow of water below the blockage.[19]

Patches of Rock

Finding a patch of fieldstone, crushed gravel, or creek rock in a depression may show where a wetland was located. The rock can indicate the inlet for a drain constructed to carry water from the deepest part of the wetland straight down into the ground, or downhill to a creek, river, or ditch some distance away. Coils of plastic drain lines were often buried beneath the rock to collect water in order to dry an area for cultivation.

Pumping Stations

The presence of a pumping station provides strong evidence that a wetland occurred on the site. Pumps are used to remove waters from a wetland where no other outlet is available that would use gravity to move water from the location. The pump is placed in a sump that collects water from a system of ditches and buried drain lines. It then moves the water to a ditch or stream farther away. An electric line and power pole are often located near the pump and its collection basin. Many pumps were first installed by a drainage district office years ago and are now maintained by the landowner who receives the greatest benefit from the system.

Looking Past the Trees

Features of drained wetlands such as open ditches, straightened streams, and dry depressions are more difficult to see once an area has grown up to trees. Looking for drained and filled wetlands in wooded areas will greatly increase the number of projects available to the land manager. From 1982 to 2002, the quantities of forestland in the United States increased by 1.9 million acres, while the acreage in cropland decreased by 52 million acres.[20] Comparing old aerial photographs with more recent photographs often shows an even greater increase in forestland in regions throughout the East, such as on the Daniel Boone National Forest in Kentucky. For example, an analysis of three areas of public and private lands within the Daniel Boone National Forest in Menifee and Rowan Counties found that open-farmed acreage changed from 27 percent in 1939 to 6 percent in 1993. Since the majority of wetlands were originally drained to create cropland, and now large numbers of these fields have grown up to trees, it is reasonable to include forested areas when attempting to identify wetland restoration needs.

Soils

While the occurrence of hydric soils is used to help delineate wetlands, their presence is clearly not a requirement for the identification of drained wetlands. The *Federal Register* defines a hydric soil as "a soil that formed under conditions of saturation, flooding or ponding long enough during the growing season to develop anaerobic conditions in the upper part."[21] Hydric soils are somewhat poorly drained to poorly drained with a water table at the surface. They may also be ponded or seasonally flooded for long duration or very long duration during the growing season.[22] Since drained wetlands no longer have standing water or a water table near the surface, they cannot meet the government's definition of having hydric soils.

Hydric soils are generally associated with wetlands and the presence of hydrophytic vegetation. Basically, unless a location has been recently plowed or disturbed by heavy equipment, it would be expected to have hydric plants growing in hydric soils because hydric vegetation is adapted for growth in soils with anaerobic conditions in the active root zone.[23]

Color can be used to help identify hydric soils during a time of year that water is not ponded on the surface or when the water table is seasonally low. Mineral soils that are hydric are generally gray colored and may contain mottles, also known as redoximorphic features. Soils that are hydric are low in ferric iron (Fe^{3+}), gray colored, and are shown in the "gleyed" color pages on the Munsell Soil Chart.[24] Nongleyed mineral soils contain a matrix chroma of 2 or less if mottled, and 1 or less if unmottled.[25]

Soils that are formed by groundwater are permanently waterlogged, with grayish colored subsoil, and their surface horizon is generally brown in color with gray mottles. Gley soils formed by surface water can be expected to have a gray-colored, seasonally saturated surface horizon, with brown-colored, aerated subsoil. Seasonally wet soils are generally a combination of brown and red colors due to the presence of ferric iron oxides along with gray, green, blue, or black colors because of the existence of ferrous iron (Fe^{2+}).[26]

Standing water removes ferric iron from soils, leaving behind the original matrix color of soil, which is usually gray. J. L. Richardson and M. J. Vepraskas describe how gray soils will turn brown again when ferrous iron moves back onto soil particle surfaces and oxidize, effectively painting them with ferric iron. This ferric iron that attaches itself to soil particles and covers them is what gives soils their characteristic brown color. Redoximorphic features can form when portions of saturated soils dry periodically. The drying

Pumping station used to drain a wetland in Kandiyohi County, Minnesota.

allows oxygen to enter soils, often by following root channels into the ground, and can cause small patches of soil to change color from gray to red.[27]

Even though it can take decades for gley soils to develop, the reddish-brown color of ferric hydroxide mottles can form within days after iron-reduced soil is aerated. During soil aeration, exchangeable ferrous iron is oxidized to essentially insoluble ferric oxide, which turns soils brown. The return of ferric iron has been found to be cyclic in seasonally waterlogged soils.[28] Drainage is done to aerate soils for the successful growth of crops such as wheat and corn. Soils are rapidly aerated by ditching, installation of buried drain lines, and repeated plowing.

Ferrous iron compounds are very highly mobile in soils, compared with ferric compounds. Ferrous iron is readily translocated and can cause mottling under conditions involving periodic flooding and fluctuating groundwater.[29]

Since ferrous iron is highly soluble, it can readily diffuse and return to aerated soil from moving water, as in runoff and groundwater. A number of examples are given by soil experts on how ferrous iron can quickly oxidize, sometimes resulting in a visible color change within 30 minutes of exposure to air.[30] When samples of saturated gleyed soils from the bottom of wetlands are collected and placed in small plastic containers already holding some air, within a couple of days the exposed surfaces on these soils have turned so brown that students have to dig into the center of the sample in order to see its previous gray color.

In *Wetlands Losses in the United States, 1780's to 1980's,* Thomas Dahl states that data on hydric soils have been used in some instances to approximate wetland acreage in the United States: "Because soil characteristics change slowly, even following drainage, summation of the soil acreages indicative of wetland conditions should approximate the wetland acreage at the time of settlement."[31] No further definition is given to the word "slowly," which has led many to believe that drained wetlands will retain a gleyed appearance and mottles for hundreds of years. This belief that soils will only change color after drainage at a slow, almost immeasurable rate has been adopted by some professionals to the point that they will only consider locations eligible for wetland restoration if they still contain hydric soils. Unfortunately, using hydric soils as a criterion for the selection of restoration sites will exclude the majority of successfully converted wetlands, as it is only in those wetlands that were poorly or partially drained where hydric soils remain.

Another point to consider is that not all wetlands contain hydric soils. Most ephemeral wetlands do not develop hydric soils, and since ephemeral wetlands were often the first to be eliminated from a landscape, looking for hydric soils to indicate the presence of drained ephemeral wetlands is not logical.

Research has shown that when rice paddies are drained before harvest, redox potentials rise; Fe^{2+} and Mn^{2+} concentrations decrease; and nitrogen, sulfur, and carbon oxidize. When soils are flooded again, the reactions reverse.[32] However, the gray colors formed in rice paddies due to wetness sometimes persist in soils long after they are naturally or artificially drained. This condition may occur where soils are not fully aerated, or where a source of ferrous iron is not available to move back into soils. It has been found that in some drained wetlands the subsoils are dull, with low chromas on the Munsell Soil Chart in

the 2–4 range. This may be caused by ferrous iron not moving back into the soils by runoff or from groundwater after aeration by draining.[33]

I have found that the soils in a drained wetland are generally brown in color without mottles. I have asked John D. Smith, Richard Bond, George McClure, and James Flowers, all lifelong drainage contractors, to describe the typical color of soil when they begin a drainage project. Their reply: gray to green. I then asked each to describe what color the soil would be if they returned 10 years later, providing they accomplished an effective drainage job. They answered unanimously, brown. They tell me that if the gray color remains, along with the strong smell, the ground is still too wet to farm and needs further drainage.

Looking at the Neighboring Land

While one farm can have absolutely no wetlands, neighboring farms may have any number of them. This situation can be due to how often a landowner participated in government-sponsored drainage programs. Richard Bond made this clear to me as we looked over some of the work he accomplished over a 50-year career as a drainage contractor in eastern Kentucky. We had just left a farm that he had worked on many times over the years when I noticed an emergent wetland in a pasture field next door. Richard said that he had never installed any drainage lines on that farm and that what we were seeing was a shame, as the wetland could have been made just as productive as the field we had just looked over, had the landowner been willing to work with the NRCS.[34]

Finding signs of wetland drainage clearly shows that wetlands were once present in an area. Unfortunately, it is not easy to determine exactly what type of wetland was present on a site without firsthand information. Fortunately, examining the shape of an area can provide clues to how deep the water might have been in a wetland. Finding a dry depression with noticeable high banks indicates the deeper waters of an emergent wetland or even a lake. On areas with less than a 1 percent slope, there may have been emergent, ephemeral, shrub-scrub, or even forested wetlands. Sites with more than a 1 percent slope may have contained small ephemeral wetlands or even the larger wet meadow, shrub-scrub, or forested wetlands.

Prior Conversion

Under the 1985 Farm Bill, the NRCS was directed to make several determinations before lands could be enrolled in USDA assistance programs. Two of these concerned the presence of both erodable soils and wetlands. Areas that had been drained and farmed before passage of the bill were to be grandfathered into the program and were labeled "PC," indicating prior conversion. The designation of PC allowed participants to receive government payments for that field, along with approval to maintain and complete further drainage actions, without loss of benefits.

The PC label was of great legal and financial importance to farmers. It allowed them to further drain wetlands without having to obtain the permits required for similar activities under Section 404 of the Clean Water Act. The PC designation protected farmers from the well-publicized stiff fines and expensive restoration actions required of those caught filling wetlands without these permits. More importantly, throughout the South, the PC designation allowed farmers to maintain government price supports for the tobacco they sold, which for generations was their number one cash crop.

The NRCS typically decided a field's eligibility by examining soil survey maps to determine if the soils were hydric or contained hydric inclusions and by looking for signs of past drainage action and crop production. If both conditions were met, the field was usually earmarked PC.

NRCS District Conservationist Marty Mc-Cleese related to me what a PC label on a map tells him today:

The field had been used to raise crops before the
    1985 Farm Bill was passed.

The field contained hydric soil or inclusions of
    hydric soils when examined.
There was evidence of past drainage activity.
The owner was participating in a USDA pro-
    gram when the field was examined.[35]

The NRCS did not make a PC determination for all fields and lands that an applicant owned, but only for those individual fields qualifying for USDA programs. As an example: A farmer owned 300 acres of land with 30 acres of fields; 10 acres of the fields were in pasture, and 20 acres were hay. The farmer wanted to start growing tobacco in 5 acres of the hayfields and to receive price support for the tobacco to be grown. The NRCS granted a PC determination for the 5-acre field where tobacco would be grown after finding it met the requirements of the 1985 Farm Bill. A PC assignment was not made on the remaining 25 acres of fields still being kept in hay and pasture, as these crops were not eligible for USDA payments at the time. The farmer could not receive government price supports to sell more tobacco than what could be grown on the 5-acre field, as the pounds of tobacco he could sell off his farm, known as the tobacco base, had been established by the government and tied to his deed years ago.

Some contractors now use the PC designation to identify drained wetlands and to justify commercial wetland mitigation sites with the Army Corps of Engineers. This approach should be used with caution because of the many drained and filled wetlands it misses. The validity of this identification procedure is weakened even more when taking into consideration how many farmers did not participate in USDA programs so that a PC label was never given to any of their drained wetlands.

Information on PC designations for farms is not generally available to the public. The landowner alone has access to this labeling, along with any government representatives working with the owner. The NRCS will not share this information with family members, neighbors, or others without the landowner's permission.

Building and Restoration

When I begin a wetland restoration workshop, I like to ask participants how one can possibly eat an elephant, the answer being "one bite at a time." The same is true for wetland restoration. Begin with one small site, not an entire landscape, and don't attempt to change the world.

Experts tell us in journals and magazine articles that we really don't know enough about wetlands to restore them and that wetland creation is a poor substitute for the real thing. Many of the young natural resource managers I meet are so worried about failing in their first attempt to build a wetland that they get talked out of even undertaking a project. I wish to challenge the views of some people, many of whom have much more education than I, by saying now is the time to build wetlands. The downward trend in wetland acreage will never be halted until more people begin building wetlands. I have yet to meet anyone with a Ph.D. who actually drained a wetland, so why do so many believe that you need an advanced degree to build one?

I believe that a barrier exists in our society that greatly restricts the future of wetland restoration. Wetlands are built with heavy equipment, and heavy-equipment operators generally learn how to run their machines from a friend or relative; only a few attend vocational school. These folks are the ones who actually build wetlands, often under the supervision of contractors who, like them, enjoy working outdoors. In contrast, the lack of success in the wetlands they build is typically expounded on by people who received years of formal education. These well-intentioned ecologists spend years examining the wetlands being built by laborers and generally attack their creations in publications that the builders never read. I have seen that when these two groups work together in an environment of mutual respect, one of the main barriers to successful wetland establishment is lifted, and the process of restoration proceeds at the pace necessary to benefit plants, animals, and society.

Many authors downplay the value of creating wetlands and actually discourage the practice of building them. Their reasons center largely on the failure of these projects to hold water as planned and on how artificial they can appear. Both wetland creation and wetland restoration projects can fail because of improper construction techniques, but, fortunately, both practices can also produce wetlands that consistently look and function like natural ecosystems.

I believe that wetland creation projects are frequently mislabeled when, in fact, they are wetland restorations, as it is easy to overlook the characteristics of successfully drained wetlands. For this reason alone, one should not discount the value of wetland creation because these projects may be taking place on top of the wetlands most effectively drained.

Many wetlands have been established with the goal of mitigation as directed by state and federal agencies working to enforce Section 404 of the Clean Water Act. Unfortunately, problems have been found with how these wetlands look and function, and these difficulties are well documented in the literature. Luckily, most wetlands established by individuals for the purpose of restoration look more natural and appear to function better than those built for mitigation. My observations relate strongly to how much ownership a person has in a wetland project. The narratives that follow describe in detail how individuals have established wetlands all over the United States, and how their great interest and dedication have produced some amazing results.

Wetlands can be successfully created on sites where no evidence is found that one ever existed. People seem to shy away from wetland creation as though there is something wrong with the practice. However, there are a number of reasons to consider a wetland creation project:

Expensive developments such as businesses, homes, parking lots, and roads have been built on historic wetlands, making them unavailable for restoration.

Not all landowners are willing to restore historic wetlands found on their property.

The historic presence of a wetland cannot be confirmed, yet suitable soils and groundwater are available for wetland creation.

Land is available for creating wetlands to provide specific habitats for the recovery of rare plants and animals.

Wetlands are desired at locations such as schools and visitor centers to provide study sites for environmental education programs.

Finding a Place to Build

Unfortunately, many books written on wetland restoration contain little practical information on site selection and actual construction techniques that will produce a desired hydrological regime. I have found that by following the directions contained in many of these publications, it is entirely possible to construct a wetland that fails to hold water as planned. Restoring a wetland does not have to be complicated, and the following chapters describe simple techniques that people have used to create successful wetlands under a number of different conditions.

When looking for wetland restoration sites,

it is best not to restrict the search to only those locations that have aquatic plants such as sedges and bulrushes. Areas that were effectively drained have typically lost their saturated soils and hydric plants. Finding lands that have fine-textured silt or clay soils is the foundation of most successful wetland creation and restoration projects.

Walking over a piece of the land and looking at its wetland potential can be a most enjoyable experience. It is best not to hurry and to have the eye of a detective to search for open ditches, especially near the lower end of an area and along the base of a hill. Open ditches are responsible for diverting a considerable amount of water from a unit of land and will often follow a straight path to a creek or river. Consider driving around and looking for large open ditches that look like streams in an area. It is also a good idea to examine aerial photographs and topographic maps to identify the signature straight-line features of large-scale ditching projects that accompany wetland drainage over a landscape.

Try to locate the previous landowner and ask if he or she is aware of any actions that were taken to dry the piece of ground, as this information can be of great value to returning a wetland. Often the driest, smoothest fields are those that were

Ephemeral wetland built in a small field near Meyers Fork within the Daniel Boone National Forest.

subjected to the most drainage and filling operations in the past.

Spend some time sampling soils on-site before designing a wetland project. Soil maps can help, but there is no substitute for firsthand examination, which may uncover inclusions and changed conditions. Determining soil texture and whether or not the water table is near the surface is critical to selecting a suitable project location. Most authors describing wetland restoration emphasize the importance of establishing suitable hydrology on a site, often going into great detail on how to calculate water budgets that attempt to take into account variables such as percolation, evapotranspiration, surface runoff, groundwater recharge, snowmelt, precipitation, and even ground cover. Granted, these factors do influence wetland function, but being able to calculate them is not a requirement for successful restoration.

Observing cattails and water in a ditch along the road often indicates areas where wetlands can be successfully constructed. These wet ditches actually contain small, linear-shaped wetlands formed from soils that were high in clay, were somewhat compacted, and collected enough runoff to hold water.

Deciding Between Surface Water and Groundwater

When designing a wetland, whether it is a restoration or a creation, you should decide if it will be supplied primarily by surface water or by groundwater. Your chance of success will increase greatly when you plan on building for one or the other. A surface-water wetland holds rainfall like a cereal bowl: within a depression made of packed soils that are high in clay and a dam that serves as a rim to keep waters from flowing downhill to a stream. A groundwater wetland is like an old-fashioned hand-dug well; it exposes a high water table and can be surrounded by soils high in sand or gravel. Different construction techniques are used to build surface-water and groundwater wetlands. A

soil test provides the information you will need to decide what strategy to use.

Use a posthole digger, soil auger, or shovel to dig a hole at least 3 feet deep near the center of the proposed wetland construction site. I prefer using a 1.5-inch-diameter soil auger attached to a 5-foot-long handle to test potential restoration sites, as a soil probe is difficult to use in rocky or clay soils. Watch to see if water seeps into the hole from the bottom and sides. If the hole fills partially or completely with water, or the slurp of water is heard as the auger is removed, a high water table is present, and you can build a wetland that will fill with groundwater. When building a groundwater wetland, you don't have to be concerned about soil texture, as the depression you make will simply expose water contained in the soil. However, if water enters from the top of your test hole, if the hole is dry, or if only a little water seeps in, plan on building a wetland that is designed to hold surface water.

A seasonally high water table can be more difficult to detect during dry weather, but there are signs that indicate its presence. One is to look for gray-colored soil or for mottling. Another is to allow more time for water to enter the test hole. A board can be placed over the top of the hole to prevent rain from entering, and you can return the next day to see if water is present in the hole. It is also a good idea to test the site again when conditions are wetter to see if water is nearer the surface.

Feel the soils below the topsoil layer, which is often dark colored and contains fine roots and organic material. Soil textures that are generally suitable for holding surface water are high in clay and silt. Several simple tests can be used to determine whether the clay or silt content is high enough to form a surface-water wetland. One involves making a 1-inch ball of soil, adding some water, and then attempting to form a thin ribbon that is 2 inches long or longer between thumb and forefinger that holds together before breaking. Soils that contain adequate amounts of clay are firm and take some strength to reshape. Molding them is like playing with a Tootsie Roll; it takes time and gets messy. Soils with high silt content feel like flour when rubbed, those with too much sand feel gritty, and those with just enough clay feel smooth and sticky. Some find it easier to test for clay and silt by rolling a walnut-size chunk of soil between their palms after adding a little water; being able to form a tube that is 2 inches long or longer before it breaks provides strong indication that soils on the site can be shaped and packed to hold surface water.

Not being able to dig a test hole because of too much rock indicates that soils are quite permeable and not suited for building a surface-water wetland. Seeing rock lying on the surface indicates that more is buried beneath, like the tip of an iceberg. Fortunately, one can often find areas that have less rock by simply moving to higher or lower ground.

Finding coarse-textured soils containing sand and gravel without the presence of groundwater near the surface limits opportunities for wetland construction. One should switch locations or plan to use a synthetic liner to build a surface-water wetland on such a site.

Now wait a minute. Experts say that many natural wetlands are supplied by both groundwater and surface water, so why not plan a restoration that uses both? Unfortunately, I have seen the majority of attempts to blend these two strategies fail. The failure may be because there are too many unknowns involving the presence of buried drainage structures, open ditches, variable soil textures, and uneven compaction during construction that confuses any attempt to calculate a water budget for the potential wetland. That is why it is best to plan on building either a surface-water or a groundwater wetland, and that when the wetland you construct to hold surface water is also supplemented by groundwater, it is a good thing, and vice versa.

Unfortunately, it is entirely possible to build a failed wetland when working on top of a historic wetland because not enough attention has been given to determining whether the soils had enough silt and clay to be shaped and compacted to hold water. Many wetlands built on fine-textured soils have also failed because of improper construction techniques, which are discussed in detail later in this book.

Adaptive Management

Perhaps the most challenging element in moving soil to restore a wetland is that what will be uncovered underground is almost always unknown. I've run into clay drain tiles, deep crayfish burrows, woodchuck holes, buried brush, rock piles, plastic drain lines, and layers of sand and gravel. Not responding to these unexpected features will result in failure of the wetland to hold water as planned. That is why it is most economical to pay heavy-equipment operators by the hour for their work, not by the job. Therefore, when contracting by the job, it is best to set aside from 10 to 15 percent of the award amount for changes needed while wetland construction is under way. Then, if a patch of gravel, for instance, is found that has to be removed and replaced with clay, the additional cost can be negotiated and added to the contract without delay.

Susan Stedman talks about using a strategy known as "adaptive management" when working to complete a wetland project: "Adaptive management is a technique that involves incorporating new information into all stages of a wetland project. Using adaptive management means you continuously evaluate your project in light of new information, generating ideas and making decisions about how to further refine the project."[1] I could not agree more. I have seen planning teams go to much work and expense in completing detailed land surveys, writing construction specifications, and preparing drawings worthy of framing only to turn over contract administration to some low-level technician who has ten other jobs to supervise that summer in random order and priority. A better approach is to have someone very knowledgeable about wetlands on the worksite every day during construction to answer questions from the contractors, look for buried drainage features, detect changes in soil texture, and modify techniques, so that the wetland will end up functioning as planned.

A common practice when building a wetland is to re-form a levee along the lower edge of a site to restore the natural basin. There is always a possibility that drain lines are present and buried beneath the location of the planned dam. Most drain lines are buried from 2.5 to 5 feet below the surface in agricultural fields. The lines are installed on a gradual downhill slope that often traverses the deepest part of the historic wetland. The drain lines can be 7 or more feet below the surface by the time they reach the outlet, which is often a ditch or stream. Don Hurst of the USDA Natural Resources Conservation Service (NRCS) routinely designed drainage systems where tile outlets were 7 or more feet deep at their outlet along the Red River in eastern Kentucky. Mark Lindflott of the NRCS has found clay drain tiles buried 12 feet deep in drained Iowa wetlands. He says that farmers had to bury the tiles at this depth to cut through the "lip," or rise, in a basin to remove water.[2]

Assume that buried drain lines and drainage structures are present, and work to locate them during construction. More than likely, you are not the first person to move soil on the site, especially considering the extraordinary efforts taken by landowners to drain lands. Most people are unaware of the presence of drain lines in a field. Even individuals who have owned land for years may not know of all the work done to install drain tiles in the area.

When restoring a wetland in a depression or on a level area of ground, dig a deep trench around the area where standing water and saturated soils are planned in order to locate and block buried drain lines. The trench should be wide enough for equipment to pack soils as they are replaced to form a tight core that will provide an impermeable foundation for the dam. The importance of packing even clay soils to get them to hold water was recognized by Henry French in his 1884 book, *Farm Drainage*. French described how heavy clay soils are not impervious to water and can be effectively drained by installing buried clay tiles.[3] Level areas or depressions that contain silt loam, silt-clay loam, and clay-textured soils that grow few if any hydric plants such as sedges and bulrushes are probably being kept dry by drainage structures buried deep in the ground.

Some believe that in order to restore a wetlands hydrology, it is only necessary to plug a ditch or block a buried drainage line. Regrettably, such limited action is rarely adequate to complete the job. Finding the main outlet for a drainage system can be difficult. Drainage systems with thousands of feet of lines covering acres of ground can all feed into one outlet. The outlet may exit in the bottom of a ditch or along the edge of a river. Outlets are often hidden by dense vegetation and may be covered by silt, water, leaves, and debris. However, the great amount of water pressure at the outlet generally keeps the system working long after the outlet is hidden from view.

Recently when I mentioned to Jimmy Lyons (NRCS), Lacy Jackson (NRCS), and Richard Bond (contractor)—a group that, excluding myself, has over 100 years of combined drainage experience—that many biologists simply advocate plugging a drain tile outlet to restore wetland hydrology, they all laughed good and hard. These experts have seen numerous drainage systems continue to work with broken and buried outlets, and they say that, since riverbanks tend to be sandy, water would likely pass through a constructed ditch plug that is anywhere near the drainage outlet. George McClure, with all his years of drainage experience, routinely digs out outlets for drainage systems that are still functioning, even when buried under many feet of soil.

Strive to compact the soils in the core and dam during construction. Soils on a restoration site can be porous from previous deep plowing, which is also known as subsoiling. Water can pass through or under a dam constructed of loose soils. Soils can be packed with a number of types of equipment, including a dozer, a sheep-foot roller, or even a truck. The key to compaction is building a dam of suitable textured soil with some moisture, placed in layers that are packed between tiers by running over them a number of times. A good standard to follow is that a dam should be built in layers 6 inches or less, packing each as it is laid down. Compaction explains how a rut in a gravel road can hold water or how all of the soil removed for a fence post can be tamped back in the hole after it is set in place. A compacted dam and core will hold surface waters even if the wetland lacks a watershed.

When working to restore a wetland downhill from the base of a mountain or hill, it is also important to dig several deep trenches perpendicular to the base of the hill in order to locate and block buried drainage structures. Open drainage ditches and buried drain lines were routinely placed along the base of a hill to divert runoff and to keep springs from saturating lower ground. These ditches and underground structures must be blocked to successfully reestablish the hydrology of an area.

If it is suspected that gravel may be located under a construction site, chances may be increased that it will hold water by moving the project uphill. The depth of silt loam soil is often greatest at the base of a hill and generally decreases downhill toward the creek. Most creeks have moved channels over time, like a snake gliding across the landscape. Each time the creek shifted, it deposited gravel layers that gradually became covered by soils from the surrounding hills.

Learning from Beavers

Studying beaver-built wetlands and what happens to them after beaver leave them can teach us much about wetland restoration. Beavers typically dam small streams with a low gradient. The streams they dam are generally in the upper portion of a watershed. Beaver dams are constructed of branches, logs, soil, and vegetation for blocking the flow of moving water, and beavers work daily to patch leaks or breaches with soil and debris.[4] Eventually, abandoned beaver dams can grow up

to shrubs and trees and become permanent parts of a landscape.

I have seen where beavers have created new wetlands by flooding non-wetland areas, and they have also changed existing wetlands by flooding wet meadow, ephemeral, scrub-shrub, and forested wetlands. Flooding caused by beavers and their cutting of trees near water can open up large areas to sunlight and change plant composition over many acres.

Beavers are able to maintain a pool of water in front of a dam as long as water flows and they are present to complete repairs. As long as water flowing into the pond exceeds that lost by movement through and beneath the dam, a pond is maintained that the beavers can use for feeding, food storage, transportation, and protection from enemies. Although the occasional beaver dam may be found across a seepage or seasonal flowage, these dams are often abandoned as, due to leakage, the beavers are unsuccessful at maintaining a large pond. Beavers are unable to block the subsurface flow of water, explaining why their dams are generally built to cross intermittent and low-volume perennial streams.

Beaver ponds are eventually abandoned due to predation, reduction in food supply, disease, or loss to trapping. Water can then flow over and through the dam, creating a breach that further lowers water elevations in a pond. A beaver pond may first be designated a forested wetland and then, as inundated trees die, the description changed to a deepwater pond. As the dam deteriorates and water levels drop, the site may become an emergent wetland, which can also contain portions of ephemeral wetlands, shrub swamps, wet meadows, and even bogs. What is unique is that a single beaver pond may contain a variety of wetland types at one time.

Beavers can do what we are often told not to do: dam a stream with a well-defined channel. Building a dam across a perennial stream with a high flow can be expensive and, if ever attempted, would require frequent maintenance. Flood events can create tremendous stress on a dam, and handling excess water under high flow con-

Beaver dam built across Rebel Trace Stream in Menifee County, Kentucky, breached within 1 month after beaver abandonment.

ditions would require the use of cement spillways and elaborate water-control structures, typically beyond the budget available to wetland managers. Damming of perennial streams and rivers is generally not allowed under current state or federal guidelines and, if planned, would require obtaining a number of permits.

There are actions that can be taken to help beavers continue their great work of creating wetlands. The most important involves maintaining habitat suitable for beavers in riparian areas. Research finds that streams with a gradient below 2 percent are possible candidates for beavers to dam on the North Carolina side of Great Smoky Mountains National Park.[5] This means that beavers could yet occupy many areas in the mountains.

In the past 10 years, most land and resource management plans prepared for national forests give riparian areas special protections that restrict tree-cutting and soil-disturbing activities. This direction may perpetuate conditions where riparian areas continue to succeed to conifers and rhododendron that beaver will not flood for food. People like Doreen Miller at the Nantahala National Forest in North Carolina have recognized

this problem and have initiated projects that cut conifers and rhododendron in order to return hardwoods along streams for beaver habitat.[6]

A beaver pond next to a road is a natural phenomenon many travelers don't often see, but I sympathize with the people responsible for keeping the culvert clear. The sound of running water has been found to trigger beaver dam-building behavior.[7] Beavers will expend considerable effort to prevent water from entering a culvert. We can help road maintenance personnel modify culverts so that they will continue to function even in the presence of beavers. Modifications should be designed to eliminate the sound of running water and may include keeping the inlet of the culvert completely under water, eliminating the sound of water flowing into the culvert, reducing the velocity of the water flowing into the culvert, and diffusing water flow into the culvert. Modifying culverts and water-control structures with devices known as beaver bafflers and beaver deceivers respond to the above needs by helping to maintain waterflow and reducing problems with beaver blockage.

One opinion held is that beavers should be allowed to take care of wetland creation as people have no business reshaping the land to establish wetlands. There are a number of problems with this view:

Beavers have been eliminated from many areas and may not colonize a former range because of vegetation changes.
Beavers are not always welcome, as they inundate agricultural fields, eat crops, block culverts, flood roads, warm trout streams, and cut ornamental trees.
Many natural wetlands, now drained and filled, were formed by forces other than beavers.
We have no control over where beavers make wetlands or over the size, depth, or type of wetlands they create. Such management control is often necessary to obtain desired plant and animal communities.
Beavers create wetlands on streams with a perennial flow, and many drained wetlands

occurred on sites fed by groundwater, springs, sheet runoff, or an intermittent flow.

The wetland restoration projects described in this book were completed in areas with small watersheds that were supported by groundwater, springs, sheet runoff, and intermittent streams. Techniques were generally used to reduce waterflow under dams and produce the desired hydrological regime without blocking rivers or streams.

When Government Permits Are Needed

It may be necessary to obtain permits from government agencies before legally restoring, creating, or enhancing a wetland. Some of these approvals are always necessary when federal funds are used to pay for all or part of a project. The most helpful step in determining whether or not permits are needed is to begin the process early by discussing your proposed plan with someone from the responsible agency on the phone or, better yet, in person. We have all grown used to the ease of websites and e-mails for handling much of our business these days. However, when dealing with regulations, there is nothing more effective than an honest, up-front discussion between you and the agency contact to get started in the right direction. Agency personnel probably will be able to answer any questions you may have that are not addressed clearly on the website, and they can ensure that you receive the correct applications.

Securing these permits can be an intense, lengthy, and costly process in the United States. Attend any conference where those who restore wetlands gather, and you will hear frightening stories of how altruistic projects were derailed by contractors and government employees who balk at issuing permits for working in an area that looks like an existing wetland. Delays and controversy often ensue when a permitting agency believes that a proposed project would change one type of wetland to another, such as a scrub-shrub to an emergent wetland. Unfortunately, what

is often not recognized is that natural wetlands that have been altered often continue to display hydric soils and plants, however changed. Some personnel don't readily see that an area displaying wetland characteristics may have looked very different at an earlier time. So, if, in a specific location, perhaps because of historic changes made to the area, the type of wetland once present has been changed, permits should be granted to allow for restoration of a wetland type that would have been more likely to have been present on the site.

Jim Curatolo is passionate about wetland restoration. As director of the Upper Susquehanna Coalition, he has secured significant grants for restoring wetlands in the upper portions of the Chesapeake Bay watershed. He, like so many others, was deceived by the degree landscapes could be changed by individuals who drained wetlands for raising crops. Every fall, Jim looks forward to hunting deer on the acreage he owns outside of Horseheads, New York. Year after year, he enjoys watching a small, ephemeral wetland near an overgrown field from the perch of his deer stand. Believing the wetland to be natural, Jim showed the site to me in July 2003. I observed that, in fact, the wetland was artificially created and was located in the bottom of a shallow, gradually sloped waterway constructed years ago along the base of a hill to divert water from the field. The field had now grown up to young forest, masking evidence of past drainage activity. Further examination yielded the presence of shallow, straight, open ditches spaced approximately 60 feet apart over the remainder of the old field, indicating the historic presence of a much larger wetland in the area.[8]

Showing concern about a few aquatic plants growing in a ditch is like the young boy who got excited over seeing a pile of elephant dung in the road as he walked to the circus—his father laughed and patiently explained that his exclamation was premature and should be saved for when he actually saw the elephant. The same is true if we get excited over aquatic plants growing in a ditch or a low spot in a field. What we are really

Small ephemeral wetland, looking natural, in the bottom of a shallow ditch dug years ago to help drain a much larger wetland in a now grown-up field owned by Jim Curatolo near Horseheads, New York.

Beautiful arrowhead, cattail, and buttonbush in a linear-shaped wetland that formed in the bottom of a diversion ditch used to drain a much larger wetland within the Wayne National Forest near Cadmus, Ohio.

brates that can help populate a restored wetland. Therefore, I regularly take steps to incorporate residual wetlands into restoration projects by requiring heavy-equipment operators to work around a sample of these hydric inclusions and even leave small sections of drainage ditches exposed in the completed wetland. Where it is not feasible to keep equipment out of wet areas, I remove their topsoil and later spread it in a completed wetland.

Small residual wetlands that form at the site of a completed drainage project are often a poor substitute for the original system. K. J. Babbitt and G. W. Tanner examined frog and toad use of temporary, isolated wetlands that remained 40 years after the drainage of an extensive marsh system in south central Florida. Their study site had been a large, usually permanent aquatic system similar in hydrology to the Everglades. It had been transformed into a cattle ranch containing numerous small wetlands connected to ditches, with varying hydro-periods. The extensive ditching needed resulted in a probable shift in the relative abundance of frog species to those more associated with fishless environments. However, some of the wetlands they studied were affected by water spilling over from ditches that contained predatory fish. In years of low rainfall, ditching was found to intensify drought conditions, thereby lowering the water table and leading to missed breeding opportunities for a number of species.[9]

Section 404 Permits

Any action that may affect waters of the United States, regardless of whether or not government funds are used, may require a permit to adhere to the Clean Water Act. This can include restoration and creation of wetlands. This application process is through the U.S. Army Corps of Engineers and your state division of water. In this way, they ensure your compliance with Section 404 of the Federal Water Pollution Control Act of 1972, as amended (commonly called the Clean Water Act).

seeing are leftovers from what was once a much more impressive, natural ecosystem.

Granted, the small wetlands remaining in ditches and depressions years after drainage have some value. They may be the remnants of historic systems and contain plants, animals, and inverte-

Not every wetland or stream project requires a permit. If your plan does not affect an existing stream or wetland, you probably do not need a license. Designing a wetland project on dry ground or upland is the best way to avoid this, and this type of layout can be possible in many situations. The vast majority of destroyed wetlands I have examined do not look like wetlands anymore. After successful drainage, the combination of standing water, saturated soil, and aquatic plants that constitute a wetland can no longer be found.

It is best to remember that when it comes to permits, there are two kinds of wetland—those that are jurisdictional, and those that are not. Jurisdictional wetlands have water present during part of the year, hydric soils, and hydric vegetation. All three are required to qualify. When walking around an area, if you see standing water and aquatic plants such as cattails, bulrushes, or buttonbush; if water rises near the surface of test holes; and if the color of the soil is gray or black, you may be looking at a jurisdictional wetland. Legally, a permit would be required to proceed.

The consequences of not obtaining a permit when one is needed can be most unpleasant and may include having to restore the site to its previous condition, completing off-site mitigation, and being penalized with a large fine.

If you work in partnership with a government agency to build the wetland, it will likely handle Section 404 Permit needs. If you are unsure that an area contains wetlands, *The Wetland Delineation Manual*[10] can help you determine whether portions of your project area are wetlands. This manual is available online by visiting most Army Corps of Engineer websites.

Before you assume that permits are needed, look on the web for the phone number of the nearest U.S. Army Corps of Engineers Office, Regulatory Branch, and call personnel there to discuss the project. However, before you do, it is best to prepare by having the following information ready:

1. State, county, and closest named community to where the project is located

2. Information on whether you are doing the project for enhancement or as a business venture for future mitigation credit

3. Whether or not you are working with a government agency to complete the project

4. Notes describing all signs of historic drainage activities, such as ditches, buried drain lines, lands plowing pattern, leveling, filling, and surface inlets

5. Whether the soil is described as hydric by the respective NRCS County Soil Survey Book, and if the NRCS has labeled it as "PC" (for prior conversion)

6. The boundary of your project work site drawn on a map and ready to fax if requested

Kathleen Kuna, who processes Section 404 Permits for the Army Corps, urges anyone who is considering a wetland restoration project to phone early in the process. The person with whom you talk will guide you through what is called a "Jurisdictional Determination," which decides if the Army Corps needs to review your project. I asked Kathleen how their office would respond to someone calling about a wetland restoration project being planned: "It's rare we get calls like that, so we'd really try to help." Most inquiries are from those who want to destroy wetlands.[11] Be upfront and honest about why you want the permit, as intent may trigger the need for certain kinds of information. Your wetland restoration project may be placed on a fast track for approval because of the agency's high regard for these ecosystems. Chances are the opportunity for an agency expert to work with you on a true wetland restoration project can bring job satisfaction not unlike that of a firefighter saving a life.

The information you share may be enough for the agency to determine if a permit is needed. Should they suggest you apply for a permit, you may have set the stage for approval to be granted under the authority of Nationwide Permit Number 27, which covers Stream and Wetland Restoration Activities.

There are two main advantages in designing your project so that it can qualify under Nationwide Permit 27. First, you will receive approval to complete the work faster because there will not be a need to apply for an Individual Permit (which can involve a 30-day review period). Second, expensive mitigation work will not be required.

Nationwide Permit (NWP) 27 is designed to cover

activities in waters of the United States associated with the restoration of former waters, the enhancement of degraded tidal and non-tidal wetlands and riparian areas, the creation of tidal and non-tidal wetlands and riparian areas, and the restoration and enhancement of non-tidal streams and non-tidal open water areas. . . . This NWP does not authorize the conversion of natural wetlands to another aquatic use, such as creation of waterfowl impoundments where a forested wetland previously existed. However, the NWP authorizes the relocation of non-tidal waters, including non-tidal wetlands, on the project site provided there are net gains in aquatic resource functions and values.[12]

This language in the NWP does not prevent you from changing one type of wetland to another type of wetland, for instance a wet-meadow to an emergent wetland. It does say that your project will not qualify if you change a natural wetland to another type of wetland. Therefore, it is critical that you make clear that you are not changing a natural wetland but are restoring an artificial wetland to its more natural condition. You must explain why you believe the wetlands in the project area resulted from human activities by describing the presence of historic drainage practices, like those described in the first chapters in this book.

I have heard of many wetland restoration projects that were delayed or rejected under the NWP because the petitioner did not recognize or explain how the wetlands present on the site were artificial wetlands. Do not assume that the government person or consultant helping you will recognize the signs of previous drainage activities within or near your project area. Unfortunately, these individuals may even argue with you that the wetlands in the project area are natural, so you must do your homework and be prepared.

You can also facilitate the approval process under the NWP by pointing out "a net increase in aquatic resource functions and values in the project area"[13] realized by implementing your project. These functions and values can include:

Providing habitat for endangered and threatened species
Improving habitat for migratory birds
Restoring spawning habitat for fish
Recharging groundwater
Reducing potential for flooding
Cleaning runoff before it reaches streams, rivers, and lakes
Increasing opportunities for environmental education
Creating new wildlife-viewing opportunities for the public
Controlling nonnative invasive plants
Restoring native plants

One way to demonstrate your concern in returning native plants to the wetland you are building is to let the government representatives know that you are purposely leaving portions of older, human-created wetlands intact to help supply aquatic plants and animals to the wetland being restored. This may involve leaving short sections of ditches that contain wetland features intact and small areas where drainage lines are not functioning undisturbed. However, be careful during construction to modify these features so that they do not continue to function after the project is complete.

Activities such as the creation of a "pond" or an "impoundment" are not eligible under the NWP and may require that you apply for an Individual Permit from the Army Corps; in this case, expensive mitigation may be required for approval. Describing the intent of your wetland restoration project clearly can reduce the possibility of confusion.

If your project cannot be covered by the NWP, you may be asked to apply for an Individual Permit. The U.S. Army Corps of Engineers Form 4345, Application for a Department of the Army

Permit, is used for this purpose. You are not required to hire a consultant to prepare this application, but you must provide the Army Corps with all the information they need on the form to process your application. Considerable delays can be incurred by submitting an incomplete application. Be aware that the Army Corps may verify the information you provide by field checking your project area.

If you question your ability to complete the Individual Permit application, Lisa Morris, Army Corps project manager, provides this advice: "If you can do your own taxes, you can complete the application."[14] The Army Corps approves 97 percent of the applications it receives. Do not let the paperwork stand between you and your restored wetland.[15]

Levy Prairie Wetland Project

Craig LeSchack knows firsthand how difficult it can be to obtain permits for restoring wetlands after working for years for the Florida Fish and Wildlife Conservation Commission. But the Levy Prairie Wetland Project he has been planning is topping them all. As director of conservation programs for Ducks Unlimited in four states, Craig is regularly involved with Section 404 and associated state permits. "It's unfortunate, but I really think we'd have the permit by now if we wanted to destroy the wetland," he says.[16]

Ducks Unlimited has entered into a cooperative agreement with the NRCS to complete the Levy Prairie Restoration Project near Gainesville, Florida. Back in the 1960s, the landowner's family began draining the 2,800-acre shallow lake by encircling it with a series of levees, from 6 to 8 feet high, and a pumping station to dry the huge basin. "They completed levees around the 1,000-acre eastern lobe and started work on the western lobe in the mid-1970s when the Army Corps of Engineers stopped them," says Craig. By the late 1990s, the landowner decided to enroll the huge acreage in the Wetland Reserve Program and stop pumping the wetland.

Levy Prairie has since reverted to wetland, but a change in hydrology has led to an explosion of cattails and red maples. The levee prevents water from entering the basin, while associated ditches continue to pull waters to the south, which aggravates flooding problems downstream in the community of Kanapaha Prairie.

Ducks Unlimited then designed a project to raise water levels from 6 to 18 inches in Levy Prairie, restoring the hydrology by returning waters via a number of structures to be installed in the levees. LeSchack says, "Everyone agrees we need to restore the hydrology and help the shorebirds, wading birds, and migratory waterfowl."

"We had no trouble getting a Nationwide 27 Permit from the Corps," he says. But trying to get a permit from the St. Johns River Water Management District has been another story. Florida is split into five water management districts, and each one holds considerable power. The water management district requires that the NRCS obtain an Individual Environmental Resource Permit before they can proceed, and Ducks Unlimited has been helping them with the application:

We filed the initial application almost two years ago. Since then, the Water Management District has come back to us with three RAI's [Requests for Additional Information]. Answering their questions has taken considerable time and cost a lot of money. An engineer, biologist, and consultant have been helping me with the process. We've done all sorts of hydrological modeling to answer their questions about how water levels might affect neighboring properties, because of the 15 to 20 other families who own land around Levy Prairie. We've had public meetings with these landowners and have found that some want more water and others don't. Unfortunately, it is impossible to balance all these requests while restoring hydrology to Levy Prairie.

Craig doesn't blame anyone for the delays: "I don't fault the water management district for what they're doing; they're just going by the rulebook. It's too bad the fact that the wetland once had higher water levels doesn't enter into the equation. Their rules are clear when it comes to drainage—you have to mitigate by building more wetlands in order to get a permit. Unfortunately, they're unclear for restoration." Craig's team has worked hard to answer all the water management district's questions:

Our rainfall modeling shows that no matter what we do, surrounding landowners are going to be affected by changes in water levels. Now, I'm hearing it's possible the water management district may require we get signed affidavits from all of the surrounding landowners stating it's okay that we make these changes. Everyone would have to sign for us to get the permit, and there's little chance this would happen. It would take only one landowner who refuses to sign to stop the project. I think the project is dead if they require this, but it will be the decision of the NRCS.

At the end of my eleven pages of notes taken while talking with Craig, he said, "I guess we're still in the middle of this complicated story." When asked if he had any recommendations for others, he replied:

As far as life lessons, it appears to me that people have a short memory of how bad flooding can get in these parts, and how restoring wetlands on Levy Prairie can reduce flooding downstream. Some of the tough questions about how water levels might affect landowners could have been asked months ago. But I still think we'll get the permit, hopefully in time for you to add it to the book.

He received the permit—just in time.

Section 401 Permits

You may need to obtain a Section 401 Permit from the U.S. Army Corps of Engineers and a Floodplain Construction Permit from your state division of water before restoring or creating wetlands near a stream or river. These permits are required before placing fill, changing streams, installing culverts, building dams, or creating small impoundments on lands that may be flooded,

even if only once every 100 years.

Do not let the need for this permit stop you from working near rivers or creeks. Building wetlands in a floodplain is a good thing to do, because so many wetlands once occurred in these riparian areas. There are a number of ways to determine if your project may be located in a floodplain:

Ask a representative of your state division of water or local floodplain manager.
Inquire from the landowner and neighbors if the area floods.
Look for large logs that have been deposited by floodwaters.
Find grasses that are matted down and pointing in one direction.
Locate vegetation and trash that are wrapped around the base of trees.
Notice whether grasses and plastic bags are hanging from fences, trees, and shrubs.
Find out whether the area is within the 100-year floodplain, as shown on a national flood insurance program map.

Some communities may require that a flood assessment be completed before wetlands can be constructed in a floodplain. Danny Fraley, who works for the Kentucky Division of Water, says, "Our main concern is that your project doesn't increase flooding problems for the neighbors."[17] A flood assessment may cost over $10,000 to complete due to the need to survey profiles across the floodplain and your proposed project. Using the following design features can help reduce the need to conduct a flood assessment, lessen the risk of erosion, reduce possible flood damage to your wetland project, and decrease the possibility of affecting the neighbors' property:

Avoid building dams higher than 3 feet or with slopes steeper than 5:1 that are perpendicular to the flow of floodwaters.
Lower the profile of the entire wetland by doing more excavation and less dam building.
Spread excess soil in low places that are

downslope from the wetland being built, or in low ridges that parallel the flow of floodwaters over the area.

Avoid the problems that carp and other unwanted fish will bring to periodically flooded wetlands by constructing them so they are shallow enough to dry and by installing drainpipes.

Cultural Resource Regulations

Wetland projects on federal land and those on state, county, and private lands receiving funding from the federal government must comply with the National Historic Preservation Act (NHPA) of 1966, as amended. This act established a program for the preservation of additional historic properties throughout the nation, along with other purposes, directing federal agencies to provide leadership in the preservation of the prehistoric and historic resources of the United States and the international community of nations, and in the administration of the national preservation program in partnership with states, Indian tribes, native Hawaiians, and local governments.

The office of the State Historic Preservation Officer (SHPO) in each state typically issues unique specifications for conducting fieldwork and for preparing cultural resource assessment reports to cover projects subject to compliance with Section 106 of the NHPA. These state specifications are intended to supplement the Secretary of the Interior's "Standards and Guidelines for Archaeology and Historic Preservation," which are often referred to as the Secretary's Standards.[18] Although not regulatory in nature, the Secretary's Standards are cited as a minimal level of performance in the Advisory Council on Historic Preservation's regulations.[19] The state-specific SHPO specifications are designed to assist historians, architectural historians, and archaeologists regarding the selection and use of appropriate field methodologies and for gathering technical inventory information for historic structures and archaeological properties that might be located within a federal undertaking's area of potential effects. The specifications also provide instructions for completing cultural resource assessment reports, including site descriptions, evaluations of eligibility for the National Register, analysis of project impacts, and recommendations for avoidance or other forms of mitigation of adverse effects.[20]

Basically, when federal funds or lands are involved, you need to make sure that your construction project does not adversely affect historic properties and archaeological sites containing standing structures or artifacts more than 50 years old, without prior written approval from the SHPO. Since most historic structures also contain archaeological deposits, the common practice is to assign archaeological site numbers to these remains, in recognition that their National Register eligibility is evaluated independently of the standing structure. Forest Service archaeologist Frank Bodkin states, "Things can get confusing when you consider that it is not uncommon for a historic structure to be evaluated as ineligible for inclusion in the National Register while the associated archaeological deposits are evaluated as eligible, and vice versa."[21]

Archaeologists will often complete what is called a Phase 1 Cultural Resource Survey of your proposed wetland project, which involves field examination of the Area of Potential Effects (APE) before you begin work. The APE is defined as the geographic area or areas within which an undertaking may directly or indirectly cause changes in the character or use of historic properties, if any such properties exist.[22] The APE may be different for archaeological sites and for historic structures.

The Phase 1 report makes recommendations for concurrence by the SHPO regarding whether implementing the project would adversely affect sites that are listed or eligible for listing on the National Register of Historic Places. "Adverse effect" means any alteration that diminishes those characteristics of a historic property that qualify it for inclusion in the National Register.[23] The specific process for determining adverse effect is found in 36 CFR Part 800, sec. 5[a–b].

Cultural resource surveys can become very expensive, to the point where their cost can easily exceed actual construction costs. They can take a long time to complete, and waiting 1 year for their results is not uncommon. For this reason, it is prudent to design your wetland project with the protection of cultural resources and site avoidance in mind. Employing the following actions will help reduce the time and expense it takes to comply with NHPA.

Inform the archaeologist conducting the survey that you plan to design the wetland project to avoid affecting cultural resources. Your goal should be to make it possible for the archaeologist to prepare a "No Find" report, which will save time and money.

Avoiding cultural resources is the fastest and cheapest way to have the survey completed. This means that you must be willing to drop construction locations or portions of them in order to avoid the presence of a significant cultural resource site. Anticipate this happening and have alternate work locations available for the archaeologist to examine at the time of the survey. This strategy works best on large tracts of land where additional locations are available for wetland work, so that finding these areas and moving to them quickly with the archaeologist is not a problem.

Ask the archaeologist to mark the boundary of any and all cultural resource sites found within or near your planned wetlands. Since most cultural resource sites are small, being less than 1 acre in size, you should be able to design around them. In general, it is much less expensive to move more soil, and to move it a greater distance, than to complete further archaeological analysis of the sites you wish to impact.

When planning a wetland project, it is best to mark each wetland site as close to the edge of the intended disturbance area as possible. Colored plastic ribbons hung from trees and shrubs work well for this purpose. Do not hand over a topographic map and ask for a survey of an entire watershed. "If money and time were of no concern, you could ask an archaeologist to study the en-

tire drainage system, but in reality, time is money, and the archaeological survey should be limited to the actual area of potential effects," says Frank Bodkin. "Contractors receive more money for each acre they survey, and they would be happy to charge you for documenting a greater number of acres, and, thus, record more historic and pre-historic sites that might lie outside the area of potential effects."[24]

Frank provides some good tips on how to exclude cultural resource sites from the areas you are marking for construction:

Since most habitable sites are located on higher ground, select low and swampy areas for your project.

Do not include a higher terrace when laying out a project along a river or large stream unless it is absolutely necessary. It is on these second or third terraces above the water where people lived.

Avoid including old home sites in your area. These are possibly cultural resource sites that require documentation, and there is a 50 percent chance that they are also prehistoric sites.

On ridges, ribbon areas with more than a 10-degree slope, as they are almost never found to be cultural resource sites, which are generally discovered on flat spots. The added expense of moving more soil on a sloped area is often offset by lower reporting costs.

Build access roads off to the side of a ridge and along the slope to avoid hitting cultural resource sites on flatter ridgetops.

Keep the blade up on a dozer and weave between trees when traversing equipment through the woods to spare the necessity to survey more permanent access routes or accidentally impact unrecorded sites.

Prepare high-quality maps that show your proposed project in an electronic format in order to speed the preparation of the archaeological report.

Mark each proposed project area with bright ribbons so that it can be easily found by the person conducting the survey to reduce wasted time and, thus, costs.

Expecting an archaeologist to support what you are doing is like believing a chiropractor will refer you to a medical doctor—it can happen, but it isn't likely. That is because, to someone who has dedicated his life to protecting cultural resources, the ground-disturbing actions you are proposing in order to restore a wetland often look the same as a highway project or shopping center.

Avoiding old home sites, which includes the house, barn, outbuildings, and lawn, is the simplest way to hold costs in check on a wetland project. Often marked on old county road maps or topographic maps, when available, home sites are readily identified by looking for the following features when marking the boundary of the area where ground disturbance is planned:

Chimney rock pile, cement foundation, and steps
Large flat stones that were used for footers
Raised ground with large trees and different grasses
Presence of fruit trees (apple, cherry, pear, plum)
Ornamental flowers such as daffodils, iris, and yucca
Accumulated or scattered household, garage, or barn refuse
Dark-colored soil and gravels
Utility pole

Archaeologists generally document the presence of old home sites they find adjacent to your project area and often require that their grounds are not disturbed. This stipulation is unfortunate, because following further analysis (Phase 2 testing), relatively few of these old home sites are found to be eligible for listing on the National Register of Historic Places. Bodkin says, "The more an archaeologist knows about a relatively common twentieth-century home site, typically the less likely it is to be recommended for protection." An old home site that is not eligible for listing on the National Register of Historic Places

is called an "Inventory Site." You can improve the odds of an archaeologist labeling a home site as an Inventory Site by providing him with information about the location. This may include the names of people who lived in the home, names of people who remember the home, and copies of old deeds, photographs, and even survey notes. The archaeologist will probably incorporate these details in the Cultural Resource Report, and you may not be required to protect its location.

Added expenses can be incurred when planning wetlands in a major river valley due to the potential for buried soil horizons, which may contain cultural resource sites. It is necessary, at some locations, to pay for digging deep test holes with a backhoe to find if alluvium is covering a prehistoric site. However, deep testing should not be required if prehistoric sites are known to occur on the surface at or near the same elevations where you are working.

Should you decide that you want to change a cultural resource site identified by a Phase 1 survey, you will be making a hefty financial investment. A Phase 2 Cultural Resource Survey must be completed for each site you wish to change. The purpose of the Phase 2 survey is to determine whether a site is eligible for listing on the National Register of Historic Places. You can expect to pay at least $10,000 for each site subjected to this process. If the Phase 2 survey finds that the site is eligible for listing, and you still want to proceed, a Phase 3 survey will be required, involving the excavation of the site. The price tag on each of these surveys is at least $50,000.

I find it best to become personally involved during the construction of a wetland to help ensure that any neighboring cultural resource sites are not affected by heavy equipment. I place cultural resource sites on contract maps and clearly mark them with bright plastic ribbons of a different color than those used to mark areas to be cleared. Before we begin work, I strongly warn all equipment operators not to disturb the marked areas or they may be personally liable for a fine. This attention to detail is especially important when moving equipment onto a worksite and during the clearing operation, as these periods are when unintentional violations are most likely to occur. I once encountered trouble with an archaeologist, incurring additional survey costs, when a dozer operator decided to move his machine over to some adjacent shade trees for lunch. These large trees happened to be growing on an old home site.

Endangered Species Regulations

The Endangered Species Act, passed in 1973, was designed to help prevent certain populations of plants, animals, and invertebrates from becoming extinct. Such species are listed as endangered or threatened by the U.S. Fish and Wildlife Service. Areas judged to be important to their survival, called "critical habitat," may also be designated by the agency.

The U.S. Fish and Wildlife Service has the responsibility of ensuring that actions funded in whole or in part by the federal government, including wetland and stream restoration, will not harm listed species or negatively affect their habitat. They do this by issuing a letter to the person proposing the project, which states that the described action is not expected to adversely affect listed species or their habitat. However, in order for the U.S. Fish and Wildlife Service to make this determination, information must be made available to them. This is generally documented in a biological assessment, a report prepared by the proponent.

The U.S. Fish and Wildlife Service is eager to help with most wetland projects, especially since the majority can benefit endangered and threatened species. First, as over 50 percent of listed species use wetlands, it is often possible to design a project to profit one or more endangered species. Second, the U.S. Fish and Wildlife Service, like all other federal agencies, has been directed to reverse the downward trend in wetland acreage, so undoubtedly, there you will find an ally in your project.[25]

"We want to help," says Mike Armstrong, con-sultation biologist for the U.S. Fish and Wildlife Service: "It's not because we have to—for most of us, it's why we became a biologist." Mike says that the time for a project manager to call him is not just before the dozer is ready to start work, when it looks like endangered species might be a problem: "It's best to talk with us before you complete the design so that we can help determine if any listed species are present in the area, and take the measures necessary, if any, to avoid impacting those species."[26] The agency manages a number of programs that are intended to encourage private landowners to build wetlands. You may find that they will prepare the biological assessment for you at no charge to help complete the project.

To help you prepare a biological assessment, the U.S. Fish and Wildlife Service can provide you with a list of endangered and threatened species found in your area. It may even be able to send you a copy of a biological assessment prepared for a similar project to serve as a guide. Recognize that you do not have to prepare a document that looks as professional as those produced by consulting companies that include an assortment of technical maps, photos, and graphs. These amenities are expensive and unnecessary. What the agency is really concerned about is how you plan to protect listed species during construction. This is why it is best to discuss your project with a U.S. Fish and Wildlife Service consultation biologist or a private lands biologist *before* you prepare a biological assessment.

It should be entirely possible for you to design a wetland project that will provide habitat that may be used by listed species and also ensure that the construction of these wetlands does not harm listed species that may be currently using the area as a habitat. The following examples show how this can be done:

The endangered Indiana bat uses wetlands on mountain ridgetops along the Cumberland Plateau for drinking. These water sources are important for Indiana bats, especially pregnant and nursing females. However, the construction of a wetland can involve pushing down trees with loose bark or broken branches that may be used by the Indiana bat for roosting. To avoid harming individual Indiana bats, the project may be designed so that these potential roosting trees are left standing or are cut down in the winter when the bats are hibernating.

The endangered Cumberland elktoe mussel, like other rare mussels, depends on clean water for its survival and is found in relatively few rivers in Appalachia. The species' survival is threatened, in part, by pollutants carried in runoff from agricultural fields, construction sites, sewage systems, parking lots, and coal mines. Building wetlands in watersheds that contain populations of the elktoe can be a good thing because these wetlands can be designed to capture and help clean contaminated runoff. Using state-approved best management practices to reduce erosion when constructing the wetland, such as sow-ing wheat, installing silt fences, and mulching with straw, can reduce the chance of directly affecting the elktoe should a sudden summer storm hit that would otherwise carry tons of sediment downstream, possibly smothering the species.

Granted, working to obtain the required approvals from the government may seem to be an overwhelming task, but it is well worth the effort. With modern equipment and techniques, it appears possible to build a wetland that can function for thousands of years. Such restored ecosystems can be expected to provide habitat for countless animal and plant species over the life of their existence. Waiting until we learn more about wetlands before we construct them will not reverse the decline that wetland-dependent rare species are experiencing but will only serve to forgo the benefits future generations may gain from our actions.

Building Wetlands on Dry Land

When building a wetland, many are quick to pass by higher areas of dry ground to look for low wet sites. Well-drained locations are often thought to be unsuitable for wetland restoration because they lack the puddles and aquatic plants believed necessary for constructing successful wetlands. Fortunately, these stories show how naturally appearing wetlands designed to hold surface water can be built on dry sites that were previously thought to be best suited for crops.

Small Fields on the Daniel Boone National Forest

On the surface, the small managed fields along Clearfork Creek in Rowan County appear well suited for wetland construction. Considering they are surrounded by small mountains, each is relatively level, with slopes ranging from 1 to 4 percent. Their soil texture is silt loam, and topsoil depth ranges from 4 to 6 inches. Theoretically, it should be easy to construct a wetland in this valley. After all, most of the natural wetlands in the region occur on silt loam soils. However, buried below the silt loam is a gravel layer, and that can mean trouble for wetland work.

A walk down to Clearfork Creek reveals a creek bed full of gravel. An examination of the exposed creek bank discloses that silt loam overlays the gravel. The practice of viewing an exposed face of a vertical creek bank to indicate the orienta-tion of soil layers uphill was also used in the late 1700s to plan drainage systems.[1] Since the 1980s, many have tried unsuccessfully to build wetlands on sites where fine-textured soils lay on top of gravel. I have seen these attempts fail in Indiana, Kentucky, Maryland, Minnesota, and Tennessee, and there are undoubtedly more.

What happens is that crayfish burrows extend down from the surface to end in water, and that water can be found in a rock layer, which can form an underground stream. Water flows from the surface down into the burrow, reaches the rock layer, and then follows the rock layer under a con-structed dam, leaving the wetland dry. Burrow-ing crayfish are often present before construction, and they will reopen their burrows even if the burrows are blocked during earthmoving. The 2-inch-diameter hole they dig can carry a lot of water into the ground. Therefore, the gravel layer must be interrupted for the wetland to develop its planned hydrological pattern.

There are a number of ways to determine if gravel is beneath a wetland project. The most reli-able is to use an excavator to dig test holes along the lower edge of the construction site. Each test hole should extend down to an impermeable layer of rock or clay, and it is best to dig as deeply as the excavator possibly can to make sure there is not a second layer of gravel. If gravel is present, it looks like washed creek rock, contains lots of water, and tends to cave in when exposed. Another way to identify a buried gravel layer is to look at soil

Ten-year-old constructed emergent wetland near Clearfork Creek within the Daniel Boone National Forest.

maps; however, gravel layers were only recorded if the soil surveyor dug deep test holes in the area. Perhaps the best indicator is to walk along the creek and locate an exposed vertical bank to see if gravel layers are present.

In the Clearfork area, since 1995, I have been the project manager overseeing the construction of thirty emergent and ephemeral wetlands from silt loam soils that rest on a gravel layer, which, in turn, overlay a bedrock layer. The emergent wetlands average 1 acre in size, with the ephemeral wetlands being less than 0.25 acre each. The following technique was used, as described in actual contract specifications, to build wetlands at Clearfork and at other areas with buried gravel layers:

The work involves building wetlands on the Daniel Boone National Forest in Rowan County, Kentucky. The contractor will complete the heavy equipment work necessary to construct the wetlands, while the Forest Service will pay for and complete seeding and planting the new wetlands. Shallow water wetlands will be built at locations shown by the attached map(s) [not included here] and as marked on the ground. Work sites are primarily open fields that have been mowed on a regular basis; however, some are old fields that have grown up to small diameter trees. Dam elevations on the wetland sites range from 1.5 to 5.0 feet high. The highest portion of each dam generally occurs in short sections that cross drainage ditches.

Contractor will furnish equipment and operators. Contractor will pay for all operating supplies and repairs. Work may begin as early as August 1, however, no later than August 15. The contract must be completed by September 30. All equipment will be located at the worksite and be ready for use as needed. The Forest Service representative will determine when conditions are suitable for work. The contractor may operate on the job site up to 10 hours per day, Monday through Saturday.

The contractor will complete the work in a man-

ner that reduces the potential for erosion. Therefore, the construction of temporary diversion ditches may be required when rain is expected in order to reduce runoff during construction. Only one work site may be opened and developed at any one time without approval from the Forest Service representative. An exception will be made on those areas where soils are too wet for construction. These locations may be ditched and drained prior to removal of vegetation and topsoil while another work site is being operated.

On Friday of each week, the contractor and Forest Service representative will agree on a work schedule for the next week. This schedule will be designed to make the most efficient use of time and equipment, based on weather predictions and existing site conditions.

The type of equipment required for this job will be two dozers and an excavator. The dozers will be equivalent in weight and horsepower to a Case 1150E; the excavator will be equivalent in weight and horsepower to a Case CX 160. The bucket on the excavator must be 48 inches wide or wider.

Required actions

1. Remove trees and shrubs from an area approximately 200 feet wide for the dam and soil for the dam as marked on the ground by plastic flags. Roll the trees cleared to remove the excess soil from their roots. Pile the cleared trees, shrubs and debris in locations adjacent to the work site as designated by the contract representative. All trees cleared that are greater than 10 inches in diameter will be set aside to be placed back into the wetland after it is built.

2. Remove vegetation and topsoil from the cleared area (approximately 175 feet wide) in front of the planned dam location as designated by the Forest Service representative. The topsoil will be stripped to a depth so that a majority of the roots are removed. The Forest Service representative will determine when the topsoil layer has been removed to a sufficient depth. Pile approximately one-half of the vegetation and topsoil in a row in front of the flags marking the downslope edge of the dam. This material will later be used to help build the backside of the dam.

3. Remove the other half of the vegetation and topsoil and move uphill, piling along the upper edge of the stripped area. This material will later be re-spread over the bottom of the wetland after the dam is built.

4. Use the dozers and excavator to make a core trench beneath the center of the dam's location, marked by flags placed by the Forest Service representative (after vegetation and topsoil have been removed). The core trench will be dug wherever the dam is built, even if the dam is only 1 inch high. The purpose of constructing the core is to prevent water from leaking under the dam. Note: the core trench will only be dug when a Forest Service representative is on site.

5. The core trench will start on high ground where the dam begins, will continue the entire distance of the dam, and will end where the dam again ties into high ground. (Imagine the wetland as a circle, with the core trench following a distance around the lower two-thirds of the circle.) The core trench will be the width of a dozer blade, approximately 12 feet wide, and will extend down to bedrock or an impermeable layer, as determined by the Forest Service representative. The core must reach below the bottom of crayfish holes and buried drain lines. On the average, the contractor can expect to dig a core trench that is up to 9 feet deep. Should it become necessary to dig greater than 9 feet deep to build the core, the contractor will receive additional compensation for the work. Compensation will be provided according to a previously established hourly rate for each piece of equipment. The Forest Service representative and the contractor will agree on the number of hours to be paid for the additional work at the end of each workday. Soil removed from the core trench that is suitable for replacement in the core will be deposited along the edge of the trench, inside of the wetland. Soil removed from the trench that is not suitable for placement in the core, such as gravel or sand, will be deposited along the edge of the trench, outside of the wetland, and can later be used to build the backside of the dam. The Forest Service representative will make the

determination of which soil is suitable for use in the core and dam.

6. Additional soil needed to build the core and dam will be taken from inside the designated pool area of each wetland, from the area previously stripped of vegetation and topsoil. This soil will be removed starting from the back edge of the cleared area and proceeding over the future pool area to maintain gradual slopes and to maximize the surface area of water in the new wetland.

7. Soil in the core and dam will be packed with a dozer with lifts no greater than 6 inches thick by running over each layer the number of times determined by the Forest Service representative until suitable compaction is obtained.

8. Where new wetlands are being built next to each other, or where wetlands are being constructed or restored next to existing wetlands, the cores of each wetland will be joined unless so stated by the Forest Service representative.

9. The top of the wetland dam will be the width of one dozer blade, approximately 12 feet wide. The inside slope of the dam will be 10:1 or more gradual, the backside of the dam will be a slope of 5:1 or more gradual. The dam will be built to the elevation of the benchmark designated at each site, generally a nail placed in a tree by the Forest Service representative. The top of the dam will be made level and be within (± 0.1 feet) of the elevation of the benchmark.

10. A spillway will be constructed on each wetland to an elevation of 0.5 feet below the top of the dam, at one end of the dam, so that water will flow gradually out of the wetland over undisturbed soils. The location and shape of the spillway will be designated by the Forest Service representative.

11. The contractor will install a Schedule 35 SDR PVC gasketed drainpipe in the wetland, placed so that it removes water from the completed wetland. Soils will be packed around the drainpipe to prevent seepage. The bottom of the wetland will be shaped so that the drainpipe can remove waters, much like a bathtub. A water control structure will be attached to the drainpipe, with assistance from the Forest Service

representative. The drainpipe and water control structure will be provided and delivered on site by the Forest Service representative at no charge to the contractor.

12. The topsoil that was previously saved inside the wetland will be spread over the bottom of the completed wetland and up the inside slope of the dam. The topsoil will be spread smoothly enough so that the bottom could be mowed or planted by a tractor when the wetland is drained. At least two large trees will be placed horizontally in the bottom of each wetland at locations designated by the Forest Service representative.

Table 12.1 shows the bid form and prices we received for heavy equipment to build eight wetlands at Clearfork in the summer of 2004.

My experience finds that it is well worth the time and money to have a qualified inspector present at the wetland project site whenever a contractor is working. When moving that much soil, the temptation for a contractor to take shortcuts is great in order to save money. The easiest time-saver is not removing all the gravel from the core, resulting in a failed wetland. It is impossible

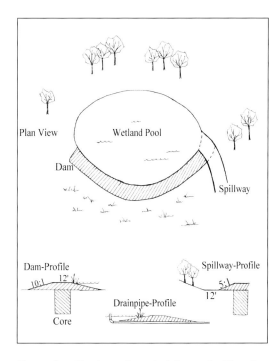

Plan and profile views of a typical dam and spillway for an emergent wetland. (Drawing by Dee Biebighauser)

Table 12.1   Bid Form and Prices for Heavy Equipment to Build Eight Wetlands, Clearfork, 2004

Clearfork Emergent Wetland Construction Information

	Wetland Number							
	1	*2*	*3*	*4*	*5*	*6*	*7*	*8*
Length of dam and core (feet)	540	720	710	240	470	408	350	478
Average height of dam (feet)	4.0	3.4	3.8	2.8	3.8	2.4	2.5	2.5
Maximum dam height (feet)	5.0	4.5	4.4	3.9	5.2	3.2	3.4	4.8
Size (acres)	0.8	2.3	3.4	1.2	1.0	0.8	1.0	1.0
Dam (cubic yards)	2,395	2,468	2,896	611	1,918	827	753	1,022
Core (cubic yards)	2,160	2,880	2,840	960	1,880	1,632	1,400	1,912
Topsoil (cubic yards)	440	1,246	1,817	672	441	527	515	503
Awarded price ($)	5,000	5,000	5,000	3,000	3,000	3,000	3,750	4,500

Note: Numbers are approximate; sites should be visited and examined by the prospective bidder to obtain the most accurate information. Please consider that portions of the yardage will need to be moved twice.

to determine whether gravel was left in the core once everything is buried. Attempting to repair a poor core is both expensive and time consuming.

Carefully follow the construction techniques described in this book, even if you decide to seek the assistance of a government agency in building a wetland. Agency representatives, though well intentioned, may be unfamiliar with the earthmoving techniques required to build a successful wetland. In fact, it is uncommon to find a wetland ecologist who has been involved in the day-to-day supervision of a wetland construction project. Over the years, I have encountered many landowners who were assured by agency representatives and contractors that it was unnecessary to go to the expense of removing topsoil and preparing a deep, compacted core for the wetland. Unfortunately, these are the ones who are now looking at drylands rather than wetlands.

Large Fields on the Wayne National Forest

From the mid-1980s to the early 1990s, the Forest Service purchased a number of farms whose acreage was added to the Wayne National Forest in Ohio. The farms were large for southern Ohio, often exceeding 200 acres. They included large bottomland fields that had been planted to corn and soybeans for generations, as well as ridgetops grown up to hardwood trees. Eventually, the acquisitions became controversial with county and township leaders, who were upset that the once-productive lands were no longer being used for any visible purpose. The Forest Service looked for ways to defuse the controversy and began allowing crops to be raised in the fields again, but only under a Special Use Permit Agreement with local farmers. Because of these agreements, corn and hay were again seen in a number of the fields, and this action was believed to have helped reduce the shock of changed ownership.

The Forest Service began holding meetings with agency personnel representing the Departments of Timber, Fisheries, Hydrology, Recreation, Soils, and Wildlife Resources to identify how the recent land acquisitions should be managed over the next 10 years. It was at these meetings that Kathy Flegel, wildlife biologist in the Ironton Ranger District, planted the seeds for what would become a large and successful wetland restoration program within the Wayne National Forest.

Kathy began by looking at a wetland restoration site near Pine Creek in Lawrence County. The large field had poor access and was growing up to trees. Not being familiar with wetland work, she asked an engineering company (already contracted to help the national forest district) to

Prospective site of an emergent wetland found in this old field with a buried gravel layer, near Clearfork Creek in Rowan County, Kentucky.

Dozer removing vegetation and topsoil from the dam location and the area in front of where the dam will be built.

Digging a core trench down to an impermeable bedrock layer, as shown by the rock dust. Loose rock is placed on the backside of the trench; silt loam is placed along the inside of the trench.

Scraping loose rock from the bottom of the core.

Removing loose rock from the core by the excavator.

The core is dug beneath the entire location of the dam.

Placing and packing silt loam in the core in layers.

Filling and packing the core with silt loam to form a foundation for the dam.

Dozers at the base of a hill pushing soil for the core and dam.

The dam built on top of the completed core.

Dozer placing gradual slopes on the front and back of the dam.

Drainpipe installed beneath the dam.

Placing a layer of topsoil inside the completed wetland.

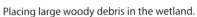

Placing large woody debris in the wetland.

examine the area to determine its wetland potential. "They said it wasn't wet enough to make a wetland. That really made me mad," she said. Not willing to accept their findings, Kathy sought other advice. "I'd heard that you were building wetlands on the Daniel Boone so I decided to give you a call."[2] I was pleased to hear from Kathy. The last time we had talked was back in 1984 when we worked together at the Superior National Forest in Minnesota.

Kathy and I met to examine the Pine Creek site in 1994. She had a vast area with which to work. Whereas a 5-acre field is huge in the Daniel

Boone National Forest, here we were looking at 40 acres. As we walked around, the field appeared quite even, so we set up an optical level and found that the area dropped by less than 4 feet from its upper to lower edge over a 2,000-foot distance, representing a 0.2 percent slope. We took samples with an auger and found silt-clay–textured soils. One could see evidence of surface drainage ditches constructed over the old field, and puddles of water stood in some of these shallow ditches. I told her that I thought the site had tremendous potential for building an emergent wetland.

Kathy and I worked together to prepare a proposal and budget for building a 22-acre wetland at Pine Creek, and then she went after the money. In 1995, she received financial commitments from Ducks Unlimited, Inc., the Ohio Department of Natural Resources, and the Forest Service that were sufficient to carry out the project.

Figuring out how high to build the dam on these large wetlands was always a process that involved a little engineering, some biology, and lots of discussion. One person would operate the level placed near the upper edge of the field; the other would walk around with a survey rod. After growing tired of shouting and waving our arms at each other, we learned it was better to communicate with handheld radios. We recorded elevations along the upper edge of the field and then along the lower edge. The large areas seldom dropped more than 3 feet from one end to the other. The maximum height of the dam was calculated by taking the reading from the lowest place along the downstream edge of the field and subtracting it from an average of the readings taken along the upper edge of the field, then adding 1 foot for the spillway and overflow pipe.

We placed a nail in a tree along the lower edge of the field that was the same elevation as the top of the dam to be built. The nail served as an important reference point during construction that we returned to every day to make sure the dam was being built level and at the correct elevation. We only needed to set up the level, place the survey rod on the nail, take a reading, and use that reading to measure elevations on the dam all day.

I remember that Kathy also had a professional engineering survey done for the Pine Creek Wetland Project, which came at great trouble and expense. As we examined the detailed contour map that was produced, I thought marking the dam's location and elevation would be easy. Not true. We both crashed through thorn-covered brush and spent hours getting hot, sweaty, and grumpy looking for ribbons and the little wire flags left behind. The area had grown up so much since the survey was completed that we were able to find only a couple of ribbons. Even finding those did not help, since, due to the dense brush, we could

see just a small portion of area. The survey map showed us that the area was quite level, but we already knew this.

We eventually dropped the idea of finding the survey stakes and proceeded to mark a location for the dam with ribbon, staying 100 feet or more from the creek. We then asked the dozer operators who were standing by to remove vegetation from the grown-up field for a distance of 200 feet in front of the ribbons we just hung. This was definitely the way to proceed, as we could now see far enough to mark the height of the dam. We learned that we could save ourselves much trouble and expense by completing a brief survey of each wetland site ourselves and, at the same time, marking the location and elevation of the dam. Kathy and I were able to complete the design and layout of a large wetland project in one day's time.

The dams would generally be laid out in a half-circle shape that started upstream at the base of the hill and finished downstream at the base of the hill. Diversion ditches were always present and functioning along the base of the hill. The dam would cross and block the lower end of the diversion ditches, a necessary step for keeping runoff in the wetland.

We gave great care to making a solid core beneath the dam during construction. Creating the core involved digging a deep trench beneath the location of the dam and then filling the trench with the soil removed, packing it in layers as it was replaced. The core trench was built to be as wide as the blade on the dozer and extended down to the bottom of crayfish burrows and beneath buried drain lines. Soils were generally not removed from the trench but were shifted and packed in place. The dozer would regularly cut through buried clay and plastic drain lines when digging the core, and the majority of them were still flowing. Occasionally, soils would be darker colored in the vicinity of a buried drainage line, indicating that topsoil was buried in the ditch when the line was covered. There was no way that Kathy, the equipment operators, or I could predict where the buried drain lines would be located or at what depth they would be found. Some would be buried as shallow as 3 feet, and others as deep as 7 feet below the surface. We had tried walking the creek to locate drain line outlets with no success, as the dense shrubs and steep, slippery banks made for poor visibility.

Another problem was the burrowing crayfish. Building the dams over active crayfish burrows would have likely resulted in failure. The 2-inch-diameter holes they had dug along the dam's location would be reopened if a deep core was not constructed to serve as the foundation for the new dam. We always dug the core down to the bottom of the crayfish burrows, and some of these went down as far as 9 feet. This was often quite a challenge, as the soils became quite soft at this depth, with water pumping out of the ground like minigeysers. It was common for the dozer to get stuck while making the core. We learned to keep the second dozer working nearby and to wrap a large chain around the back end of the dozer digging the core. Kathy or I always watched the dozer as it was coring and, when it got stuck, motioned to the other operator to come over and pull him out. In this way, the operators were able to stay on their machines while we handled the chain. Getting the dozer stuck and pulled out involved about a 10-minute delay, and we made sure the drivers knew their pay continued through the entire process. Had we not taken these steps, the dozer operators would have failed to go deep enough to make a good core, possibly missing drain lines and crayfish burrows that would have resulted in an unsuccessful wetland. An excavator could also have built the core, and with no danger of getting stuck; however, we were always watching dollars so closely that we never brought one in to give it a try.

The various heavy-equipment contractors who have worked to complete the wetlands within the Ironton Ranger District have all been from Kentucky and have all stayed in motels while the projects were under way. Since no one could afford sitting in a motel while a site dried, we learned how to keep from being stopped by rainy weather. Each day we worked to complete a 200- to 300-foot-long section of dam, and we only removed

topsoil from an area where we were certain that soils could be excavated and used that day. Exposing any additional areas of soil would result in a muddy mess if it rained, requiring that the site dry for several days before we could get back to work. By not disturbing ground in advance, we found that we could work the very next day after a rain event.

The Forest Service hired heavy-equipment contractors and paid them by the hour for these wetland projects. A Forest Service employee was always on site to direct operations and to assist with construction activities for every day the contractor was working. Paying by the hour had many advantages over issuing a "turn-key" construction contract with specifications. We were able to rapidly respond to changes in underground conditions by digging a deeper core, removing topsoil when needed to a greater depth, moving soils a greater distance, and changing the dam's location to avoid sandy soils when necessary. These changes could be made without incurring additional expenses and delays often associated with negotiating written change orders. However, disadvantages were readily apparent at times when

Kathy Flegel marks the dam location for the Cadmus Wetland Project in Gallia County within the Wayne National Forest.

Removing topsoil from the dam and area in front of dam for the Cadmus Wetland Project.

Digging a deep hole for the core, where the dam begins at the base of the hill. The core extends down below the crayfish burrows.

Preparing the core for a new section of dam.

Dozer extending the core along the location where the dam will be built. Topsoil was piled along the outside edge of the core.

Building the dam on top of the packed core. The topsoil will later be used to form the back slope of the dam.

Pushing soil to build the highest portion of the dam by backing up 150 feet from the dam's location. The topsoil piled behind the dozers will later be spread over the borrow area in front of the dam.

Leveling and packing the top of the dam. The cover for the water-control structure is to the left of the dozer.

Making a gradual slope on the inside of the dam.

Providing a gradual slope on the back of the dam to prevent washing under flood conditions also helps make the wetland appear more natural.

Setting the PVC water-control structure and drainpipe in place before covering with soil.

Inlet of water-control structure modified to reduce plugging by beavers.

An 8-year-old emergent wetland built by Kathy Flegel near Sandfork Creek in Gallia County within the Wayne National Forest in Ohio.

an operator appeared to hold back on the throttle and not push the machine as hard as one might normally do when getting paid by the job.

In the summer of 2004, Kathy completed her fourth wetland project, this time near the small community of Cadmus in Gallia County. "This will be my last hurrah before I retire," she said.[3] The Cadmus Project involved building a 26-acre emergent wetland and four ephemeral wetlands that ranged from 0.25 to 1 acre in size. It took two Case 1150 dozers operating for a total of 280 hours to complete the long dam, build three islands, and spread topsoil over exposed borrow areas. Another 50 hours of dozer work was used to build the ephemeral wetlands. At $75 an hour, heavy-equipment use totaled $25,000 for the project.

A drainpipe was always placed in each large wetland during construction. The drainpipe provided the ability to remove water when carp or predatory fish became established and would facilitate the repair of damage that beaver and muskrat caused to a dam.

Eddie Park, a Forest Service biological technician who worked with Kathy Flegel to maintain and manage the wetlands, has seen where beavers have blocked drainpipes leading to water-control structures we installed on earlier projects. He has since gone back and modified many of these so that beavers are less likely to plug them. Eddie's improvements involve attaching a 20-foot-long section of perforated pipe to the inlet on the drain-pipe, then surrounding the perforated pipe with a wire cage. Eddie said: "The beaver will leave it alone if they can't hear or feel water running. I did have one problem where beaver buried the cover on the water-control structure that was set in the dam. They didn't hurt anything. I guess the noise of water running must have [driven] them nuts."[4] Eddie might be right, as beavers now appear to be leaving the water-control structures alone that we modified according to his directions.

In telling this story about their wetland program, Kathy stressed the importance of maintenance. She said, "You can't just make them and walk away." She manages eight emergent wetlands totaling 124 acres, along with a dozen smaller ephemeral wetlands on the Ironton Ranger District. Maintenance involves the annual mowing of the dams with a tractor and bush hog, using a chain saw to cut trees that escape the mower, repairing occasional muskrat holes with a backhoe, modifying water-control structures so that beaver will leave them alone, replacing leaky water-control structures caused by aged gaskets, and drawing down water levels to remove rough fish such as bowfin. Including Eddie's salary, Kathy budgets from $5,000 to $6,000 a year for wetland maintenance of the ranger district.

Kathy gets excited when describing all the wildlife species using the restored wetlands. Blanchard's cricket frogs, a rare species in the Forest Service category "sensitive," breed in the wetlands. Double-crested cormorants began nesting, and great blue heron rookeries have been found. Ospreys are often seen hunting overhead, and there's an abundance of wood duck, mallard, and Canada geese nesting taking place.

Use of the Levee Plow

Pulling a levee plow behind a powerful tractor provides perhaps the fastest, most efficient way to construct levees on level land. Levee plows generally consist of eight to twenty-four disc blades joined by a sturdy, adjustable frame; they are designed to be pulled by laser-level-guided tractors of at least 150 horsepower. One pass with a levee plow results in the construction of a low dam, whose height can be increased by each subsequent pass. The outside discs of a levee plow pull soil inward to build up a levee of soil just behind the plow. Under most conditions, the highest levee that can be reasonably built with a levee plow is 3 feet. The levee plow is generally used to build dams on the contour, with a new dam being stair-stepped on every 2.5-foot change in elevation. Dams built with a levee plow are generally narrow and have rounded tops with little compaction. Levee plows are widely used to temporarily flood fields for rice production. These fields are often rotated to corn or soybeans in subsequent years. The same levee plow can be used to level an existing dam when the gangs are adjusted to turn the disks outward.

There are a number of serious disadvantages to using a levee plow for wetland construction. The plow rides over buried drain lines and crayfish burrows that can carry water out of a wetland, resulting in failure. The plow typically places organic material such as roots and plant stems into the dam, which can cause water to leak through the dam. Since construction involves little compaction, these dams should be considered suitable for temporary applications on saturated soils without buried drain lines.

Use of the Malsam Terrace Plow

Bud Malsam of WaKeeney, Kansas, developed the terrace plow early in the 1960s to help farmers terrace agricultural fields in the Midwest. His invention has since been used throughout North America to build levees for wetland establishment. Basically consisting of a large disk with a conveyor belt, the device is pulled by a tractor of at least 120 hp to pile a continuous row of soil that becomes a dam. According to Doug Malsam, Bud's son, it is entirely possible to construct a 1-foot-high levee with a base 22 feet wide that is 1 mile in length in a day with a Malsam terrace plow being pulled by a tractor at 4.5 miles per hour.[5]

A Malsam terrace plow works best when building levees 200 or more feet long that are less than 4 feet high, on somewhat level ground. Being able to pull the plow with a tractor of a size commonly owned by farmers presents an advantage to those interested in wetland work when normally a dozer would have to be hired to move soil at a greater cost.

Brad Feaster, assistant manager with the Indiana Department of Natural Resources (DNR) at the Hovey Lake Fish and Wildlife Management Area, has successfully used the terrace plow to construct levees for wetland projects in southeastern Indiana. He says that it is worth investigating using one, as a dam can be built in half the time it takes a

dozer. Brad described why the Indiana DNR purchased a terracer 5 years ago: "They're the way to go for those long, low dams, and they're cheaper to use than a scraper, but for short dams or those you need to build taller, you're better off with a dozer." He went on to explain that they generally build a dam about 2 feet higher than needed with the terracer in order to allow for settling, and that they have had to return on some occasions with a dozer to grade the dam after settling occurred.[6]

Brad provided me with detailed information on how he used the terrace plow to build the majority of levees needed to create 36- to 37-acre emergent wetlands in cooperation with Ducks Unlimited, Inc.:

Disc an area 40 feet wide on both sides of the levee's centerline to remove existing vegetation. Perform the disking as far in advance of the levee construction as possible, and as many times as needed to remove the vegetation. Vegetation removal is critical; otherwise, the residue will compromise the levee base.

I think the book on the terracer recommends a minimum 130 hp tractor with 1100 RPM PTO. We have done well using a 110 hp tractor with front wheel assist. We have also used a 90 hp front wheel assist tractor as shown in the photograph, however, the engine temperature would tend to run hot, and we would often have to let it cool down. Tractors with dual rear wheels do not perform well.

The tractor must also have three hydraulic control valves. Since we only had two sets of controls, we had to install *FASSE* valves on our tractors. The *FASSE* valves are more or less a splitter that allows one hydraulic control to operate two hydraulic cylinders through the use of a manual switch controlled by the operator. If someone was going to use the terracer on a regular basis, and did not have a tractor with three hydraulic controls, I definitely recommend retrofitting the tractor with the third control or purchasing a new tractor with three controls. The *FASSE* valves tend to leak fluid and are somewhat cumbersome for daily use. Since we do not use the terracer that often (maybe 2 weeks/year), the *FASSE* valves work well for us.

The terracer is somewhat complicated to use at first, but is easily mastered after just a few hours. It removes

Malsam terrace plow used by the Indiana Department of Natural Resources to construct a levee for a wetland at the Hovey Lake Fish and Wildlife Management Area. (Photograph by Brad Feaster, Indiana Department of Natural Resources)

from 4 to 5 inches of soil with each pass. We try and get the tops of our levees 8 feet wide with a 3:1 slope. In ideal working conditions and with an experienced operator, 2,000–3,000 feet of 3-foot-high levee can be built in a day. To date, we have only built levees by borrowing dirt from one side (the inside) of the levee, so in ideal conditions we can usually get 1,000 feet per day 3-foot high. By the time you're done, the borrow area in front of the dam has the shape of the dam, only turned upside down.

It is important to only work on a length of levee that can be completed in one day. Should it rain overnight, the borrow area fills with water and must then be drained and dried before continuing, we learned that the hard way.

Follow the terracer with a small tractor pulling an offset disc. This will help spread and compact the newly placed material. We have hired a dozer to help with the final dressing and compacting, although this is not always necessary.

Since water-control structure placement and levee integrity near the structure are critical, it was recommended to us that we contract structure installation and levee construction for at least 50 feet on either side of the structure. We have always done that.[7]

Brad had the contractor use a dozer to build approximately 2,000 feet of levee, including those portions on either side of the water-control structures. The dozer was also used to dress and compact another 600 feet of levee that contained numerous large dirt clods. Portions of the levee with a height greater than 3.5 feet were also built with the dozer, with some sections reaching 7 feet tall. "Beginning to end, we completed our portion of this project in about two weeks," stated Brad.

The terracer was pulled by a tractor for a total of 50 hours to build approximately 6,500 feet of levee with an average height of 3.5 feet for this project. A tractor was used for approximately 6 hours to prepare the site by disking and another 20 to 25 hours to disk the levee and run over it for compaction. It also took about a day's work for personnel to maintain the equipment for the project.

To protect the newly constructed levees from erosion, exposed soils were seeded with winter wheat and oats, fertilized, and mulched around the water-control structures with straw. They returned to the levees the first winter and used a tractor and drill to sow switchgrass for more long-term erosion protection.

Brad says: "We began construction in November of 2000. Winter weather, rains, and spring floods did not allow us to finish until June of 2001. It did not take long for wildlife to find it once we flooded it in July 2001. This particular wetland complex has been extraordinarily successful in attracting shorebirds, wading birds, and grassland nesting birds."

Brad has always used the terracer to borrow soil for the dam from inside the wetland area next to the dam. He says that greater efficiency can be obtained by borrowing soil from both sides of the dam, but that this would result in a smaller wetland, as one would need to stay farther from the creek. He has cut through a number of buried clay drainage lines with the terracer and found that most were still working. Some of the wetlands he has built failed to hold water because of missed drainage lines. He would sometimes find a leak on the backside of the newly constructed dam that indicated the location of a buried drain line. In these cases he returned to the wetland with an excavator to expose and remove the drain lines.

Sloped Fields

Fields with a visible slope are often overlooked for wetland establishment; however, they can be fashioned into a variety of wetland types. It is possible to shape wetlands on hillsides so that they look and function like wetlands on more level ground. National forests located in mountains were often assembled of lands that nobody wanted and included fields poorly suited for agriculture. Sometimes the only sites now available for wetland development are old fields on hillsides with slopes reaching 5 percent.

I have worked with a number of contractors to establish both emergent and ephemeral wetlands on openings with a 4 percent or greater slope within the Daniel Boone National Forest. We have developed techniques that improve the outcome of building wetlands on hillside areas and also help to keep costs reasonable.

Diversion ditches are generally found at the base of a mountain and run along the upper edge of sloped fields. Water flowing into these diversion ditches can be turned so that it enters wetlands constructed below. In addition, small intermittent streams flowing off the mountains are often channeled and straightened so they no longer flow over the field. These straightened streams, which may be acting as open ditches, often form the left and right edges of a field. These streams also can be directed to enter the constructed wetlands, providing they do not carry too much water. Streams that have a large watershed can cause water levels to vary greatly in a wetland and may threaten the integrity of constructed dams.

When building a wetland in a sloped field, the tendency is to make the dam high enough to flood the entire field. Unfortunately, such action may result in creating deep areas of water that will not grow plants and will function more like

Series of wetlands stair-stepping like rice paddies down a 4 percent slope in an old field along Beaver Creek within the Daniel Boone National Forest.

a lake than a wetland. I generally find it better to construct a series of small wetlands on a slope, stair-stepping them down the hillside like rice paddies rather than making one large wetland. A series of shorter dams is generally cheaper to construct than one tall dam, and the greater number of wetlands increases the opportunity to provide a variety of habitats in one locale. Look for terraces or benches on which to create wetlands. If benches are lacking, building dams on a contour every 4-foot change in elevation will help keep them shallow.

The dozer operator should be instructed to begin removing soil for the dam from the upper edge of the pool area where water is to reach. The operator proceeds to level a portion of the slope, reversing grade so that a shallow pool of water will form in front of the dam. I place a gradual 5:1 slope on both the inside and backside of the dam, so it will appear more natural. It is just as important to place a gradual slope on the upper edge of the wetland where soil is removed for the dam. Forming these gradual slopes helps keep the wetland from becoming too deep during construction, will make it possible to mow the dam, and promotes conditions for plant diversity. We often place a drainpipe and a spillway on each wetland, and then run the overflow from the highest wetland downhill into each successive wetland on the slope.

Higher costs associated with moving more soil on a slope can be easily offset by lower administrative costs. Because we humans rarely lived on steep slopes, there is a very low probability that cultural resources will be found on these hillsides. In addition, sloped fields are generally above the floodplain, so state floodplain construction permits and U.S. Army Corps 401 permits may not be needed.

Forested Areas

Restoration of bottomland hardwoods and forested wetlands has been receiving considerable attention throughout the southeastern United States. Often the main management action reported for bringing back these ecosystems involves planting a diversity of oak, maple, gum, and pitch pine in agricultural fields located in riparian areas. A visit to Congaree Swamp National Monument in South Carolina, which is home to the largest remaining virgin bottomland forest in the Southeast, shows that planting alone will not restore these valuable wetlands. At Congaree one also sees an abundance of shallow pools of water, most appearing to be ephemeral, surrounded by large-diameter old-growth trees. Unfortunately, depressions that seasonally contain water are often missing from bottomland hardwood restoration projects, a situation due to historic land leveling and drainage practices.

Heavy equipment is generally needed to restore the landscape and hydrology of drained bottomland hardwoods. One should suspect that any waterway, whether it is a stream or ditch, has been modified to rapidly remove water from a bottomland hardwood site. Buried drainage structures are also likely to be present, so it may be necessary to dig deep trenches across waterways in order to locate and block pipes, buried rock, and wooden drains. The lack of pools and the presence of dry depressions provide strong indications that an area has been leveled and drained. One should also examine the base of adjoining hills to identify diversion ditches, many of which have drainage structures buried in them.

I have used two main strategies to establish forested wetlands on sites with silt loam and clay soils. The first involves building a levee that will allow for the periodic flooding of already established trees; the second involves building a levee but keeping the wetland drained until trees become established on the site. Although both strategies work, the former produces faster results.

Bankhead National Forest

Tom Counts remembers deer hunting as a young boy near a wet place on top of a forested ridge near the Sipsey Wilderness in Lawrence County,

Alabama. The 0.5-acre area would have ankle-deep water in the winter and spring. Tom's love for the outdoors turned into a profession as he began a career with the NRCS as a wildlife biologist. After 22 years with the NRCS, he accepted a wildlife biologist job with the Forest Service. In 2002, he returned to the wet area he used to hunt on the national forest while reviewing a proposed trail-head parking lot for the Bankhead Ranger District. Having gained considerable knowledge of wetland ecosystems over his career, he wanted to walk around the area once again. This time, he noticed a shallow ditch carrying water from what he now recognized as a drained gum pond, a type of ephemeral wetland known for the large gum trees they contain. He had heard that the Tennessee Valley Authority once drained wetlands in the area for mosquito control, and he suspected that he might have found one of these "malaria ditches."[8]

Tom wanted to restore the wetland by filling the drainage ditch. He worked with NRCS and Forest Service personnel to complete an on-the-ground survey and design for the wetland restoration. He then sought funding from the Forest Service Soil and Water Program to restore the wetland, but his request was denied. The Land and Resource Management Plan for the Bankhead National Forest contained little direction on wetlands, and there was little support for their restoration. So, during the next fiscal year he took a different approach by changing the name of the project and applying for funding from the Forest Service Endangered Species Program. This time, he mentioned that the drained wetland was located less than a mile from a limestone cave used by the federally endangered Indiana bat. Since the Indiana bat makes extensive use of ridgetop wetlands, the project was chosen as a good way to improve habitat for the species.

The team used a low-impact method to restore the wetland. Not wanting to remove trees or disturb the site, they decided to haul in a dump truck load of clay. The sides and bottom of the old ditch were prepared for a clay seal by using fire rakes to clean out the organic material. A

(top) Restored forested wetland near the Licking River in Bath County within the Daniel Boone National Forest in 2003.

(bottom) Complex of small wetlands established in a bottomland field grown up to trees along Lower Lick Fork within the Daniel Boone National Forest. The largest wetland on the left was built with a dozer to trap runoff; the smaller wetlands were built with an excavator to expose a high water table.

tractor with a front-end loader was used to carry clay over to the ditch. Three people raked 2- to 3-inch layers of clay into the ditch at a time, packing each layer with fence posts. They proceeded in the same fashion to fill the entire 20-foot long ditch, mulching over the top of the packed soils with

Willows beginning to grow in this newly constructed wetland that has been drained each summer in Menifee County, Kentucky.

hardwood leaves and pine straw. No seeding was done, and the area appeared quite natural when finished. The restored wetland is now knee deep and about 1 acre in size.

Beaver Creek Forested Wetland

Finding a place to build a forested wetland that will inundate large trees can be a challenge. I found such a location in 1993 when searching for wetland restoration projects along Beaver Creek within the Daniel Boone National Forest. A diversion ditch approximately 1,000 feet long, 12 feet wide, and 4 feet deep had been constructed parallel to Beaver Creek to dry agricultural land. I was told by Donald Back, who used to farm the area, that this was one of many ditches the SCS helped construct in the 1950s.[9] The productive fields drained by the ditch had been planted to corn, tobacco, and hay for years until they were condemned by the Army Corps of Engineers in the late 1960s. The government acquired the lands in the event that the Cave Run Lake Reservoir, which was being constructed at the time, should reach maximum flood pool. The Army Corps

later turned the lands over to the Forest Service, who planted a number of the fields to pine and allowed the rest to grow up to hardwood trees.

Building the forested wetland was simple. All we did was construct a dam across the outlet of the diversion ditch to restore the natural levee found along the banks of the creek. To determine how high to build the dam and to mark where it would begin and end, Richard Hunter and I used a level and rod to record elevations along the ridge following Beaver Creek. We found that the ditch was 4 feet lower than the creek bank, so the dam would need to be at least 4 feet high at that point. We added another 6 inches for the spillway, then pounded a nail in a tree at an elevation the same as the top of the dam that would serve as a guide during construction.[10]

The Forest Service issued a contract to Steve Allen that paid him by the hour for the dozer work needed to complete the job. Steve brought in a Case 1150 E dozer for moving the soils. Our first step was to remove topsoil from the future dam and borrow location in front of the dam. The silt loam soil we needed to build the dam was located in the potential pool area for the wetland, on either side of the drainage ditch. Fortunately, we did not have to remove any large trees from the borrow site as it was a dense tangle of multiflora rose, an exotic invasive shrub.

After the vegetation and topsoil were removed, we cored beneath the dam's location to a depth that was under the crayfish burrows and built a dam on top of the packed core. The dam had a maximum height of 4.5 feet over the deepest part of the ditch and was 250 feet long. It was capable of flooding 5 acres of trees to an average depth of 12 inches.

We placed a drainpipe constructed of 6-inch-diameter PVC sewer and drain pipe joined with rubber gaskets beneath the dam, and installed an in-line fiberglass water-control structure with sliding panels along the drainpipe and buried in the dam. To keep vehicles off the dam, we blocked an old field access road with a pile of soil. Fall and winter rains gradually filled the wetland. The wetland is now managed so that its sycamore, willow, sweet gum, and maple trees are surrounded by shallow waters from winter to early spring. When visiting the wetland, I often flush wood ducks, mallards, and black ducks. Because the water-control structure is opened each spring to dry the site to keep from drowning the trees, the borrow area has grown up to a dense stand of sapling-sized hardwood trees.

Looking back on the project, I see that if we had wanted the dam higher and the wetland deeper, it would have required the removal of the trees along the creek bank for locating a longer dam and to obtain more soil for the dam. Some people say that the construction of a ditch plug is all that is needed to restore a wetland drained like this one. I disagree because I've seen and heard how much heavy equipment is used to reshape land and to cut through natural levees when constructing diversion ditches to drain these wetlands.

Slabcamp Diversion Ditch

Forester Ron Taylor was surprised to discover that he had created a wet meadow wetland in the forested drainage of Slabcamp Creek within the Daniel Boone National Forest. While working to reduce erosion in the riparian area in 1992, he used a small dozer to build a 3-foot-high mound of soil to block a 4-foot-wide diversion ditch located at the base of steep hill. The ditch had most likely been constructed by hand in the 1800s to help dry the narrow bottomland for farming. In 2005, we discovered that his action had restored a sheetlike pattern of runoff over a 0.33-acre spot between the hill and the creek. The soils in the water's path had become saturated and were now supporting a diversity of sedges, grasses, and ferns, which were not like the drained forest along the stream. The small wetland was surrounded by trees yet was too wet to support trees itself. Ron's act shows the ease with which some wetlands can be established and demonstrates how some projects can be accomplished in ways that avoid affecting established trees along a stream.[11]

(top) Hydrologist Tony Crump measures the depth of a 20-foot-long ditch used to drain a gum pond on the Bankhead Ranger District in the national forests in Alabama. (Photograph by Tom Counts, USDA Forest Service)

(bottom) View of the same gum pond in Lawrence County restored in 2003 by hand-packing clay into the drainage ditch. (Photograph by Tom Counts, USDA Forest Service)

Where a Scraper Can Help

Dennis Eger of the Indiana Department of Natural Resources and Pat Merchant from the USDA Forest Service had a vision for creating a forested wetland near Stinking Fork in Perry County

Forested wetland constructed in 1993 at Beaver Creek in Menifee County within the Daniel Boone National Forest.

within Indiana's Hoosier National Forest. They wanted to build a wetland that would seasonally flood large bottomland hardwood trees such as silver maple, sycamore, cottonwood, and green ash. Pat identified a potential 12-acre site on a recent land acquisition from soil maps and successful experiences building wetlands where smaller creeks like Stinking Fork enter larger rivers like the Little Blue River. The forested wetland they designed required the construction of a 1,000-foot levee with a maximum height of 5 feet. Levee construction often involves obtaining borrow from the front of the dam to keep costs reasonable, which can result in the clearing of large areas of trees and shrubs. Removing trees from the potential pool area at Stinking Fork would have been expensive and gone against the objective of seasonally flooding them for wildlife.[12]

Eger said, "The main reason we used the scraper is that it allowed us to maintain standing trees in the potential pool area for wood ducks. We left trees standing in what would be the normal borrow area for the dozer. There were open fields of silt loam and silt-clay soil on either side of the area we wanted to flood, but the soil in these fields was too far to push with the dozer." Eger was

also interested in using a scraper to avoid creating deep water next to the levee that can occur when borrow is taken from the front of the dam.

It took a D5 high track dozer and a D6 dozer 3 days to remove the 12- to 14-inch-diameter trees from the location where the levee would be built. Excavating the core trench for the dam was simply a matter of connecting the deep stump holes, said Dennis. Some of the trees removed were placed in the future pool area, but most were placed in an interconnected windrow between the dam and the creek to reduce downstream movement in a flood event.

The Forest Service rented a self-loading Cat 613 scraper capable of transporting 11 yards of soil for the job. Dennis said, "This is considered a small scraper, as we have some working in the coal mines around here that can handle 40 yards." It took a bit of planning to figure out how to move the scraper to the job site from Indianapolis, about 2.5 hours away. Forest Service Engineer Keno Kohl was a big help as he plotted the route to the job site. Keno checked weight limits on the bridges and met with Perry County engineers to obtain approval for transporting the 33,650-pound scraper on county roads. Due to weight restrictions on the bridges, they unloaded the scraper from the trailer and drove it 4 miles down the road to the job site. Dennis mentioned, "You have to watch out for right-angle turns on those gravel roads because there is no way a large tractor and trailer can make it around them."

Topsoil was removed from the borrow sites and stockpiled for later reclamation activities. The scraper was run in a figure-eight pattern to pick up clay on one end of the levee, dump clay on the levee, and then pick up clay on the opposite end, cross back over the levee dumping clay, go back to the starting point and repeat the process again; "pretty efficient, actually," declared Dennis.

Dennis explained that the most efficient way to utilize a scraper is to remove soil from pits on both sides of the levee being built. A circular pattern is followed when soil is removed from the bottom of the future wetland being built. One must avoid climbing a slope at an angle as the scraper can turn over if a turn is made; a scraper is best suited for level areas:

The key is to keep the scraper moving and to remember that it has the right of way on the construction site. Nothing should be done to slow or stop it. It is essential to grade with the scraper blade when unloading and to keep the scraper roadway as smooth as possible. The dozers are there to support the scraper by keeping the path of travel and borrow areas smooth. In extra hard dirt, the dozers can push the scraper to facilitate loading, break the dirt loose, and form windrows for easier pickup by the scraper. Smooth roadways and easy loading dirt are essential to efficient scraper operation.

They operated the scraper 5 days a week for 3 consecutive weeks. Receiving rain the fourth week, they attempted to resume work when the soils were wet. "An empty scraper doesn't get any traction when the ground is wet—it slides all over the place," said Dennis. They were able to use the scraper only after a dozer had scraped away the wet soils on top of the ground.

Indiana DNR personnel operated the heavy equipment used to complete the project. The scraper was used for approximately 135 hours over a 4-week period, the D5 dozer for about 120 hours, and the D6 for around 112 hours. The equipment operators were paid an average of $20 an hour for their work. Total cost for heavy equipment and operators was estimated to be $30,740:[13]

scraper = ($7,800 rental/month) × (1 month) +
    (135 hours labor @ $20.00/hour) =
    $10,500
dozer = ($5,200 rental/month/dozer) ×
    (1.5 months) × (2 dozers) +
    (232 hours labor @ $20.00/hour) =
    $20.240
scraper and dozer use = $30,740

Two 4-inch-diameter PVC drainpipes were installed beneath the dam so that water could be removed from the wetland to keep the trees alive. A spillway was created approximately 1 foot below the top of the dam to handle overflow. The wetland

(top) Scraper loads soil from a borrow source for use in building a dam on the Stinking Fork Wetland Project within the Hoosier National Forest. (Photograph by Dennis Eger, Indiana Department of Natural Resources)

(bottom) Recently completed forested wetland near Stinking Fork in Indiana. (Photograph by Dennis Eger, Indiana Department of Natural Resources)

levee was seeded to wheat and red clover, fertilized, limed with pellets, and mulched with straw. After excavation, one of the borrow areas was connected to become part of the wetland, and topsoil was spread over its exposed mineral soil.

Timber Harvest Units

Areas where trees have been harvested can provide excellent locations for building small wetlands. Timber sale units already have access roads in place, and large trees have been removed so there is more room to maneuver equipment. On public lands, archaeological surveys have generally been completed for timber sale units, saving the planner both time and money. Another good reason to use them is that harvested areas often look so bad that few people care if soil and vegetation are further disturbed. Disadvantages include that recently cut trees have stumps that are difficult and costly to remove, and if harvested areas were planted to trees, foresters may be reluctant to inundate their investment.

A. J. K. Calhoun and P. deMaynadier reported that the artificial pools created by skidder ruts, borrow pits, blocked road drainages, and road ditches in harvest areas often dried significantly faster than natural pools. They claimed, "Therefore, artificial pools may not replace natural pools as successful breeding habitat for wood frogs and spotted salamanders because the period of time the pool holds water is too short to permit successful juvenile recruitment (especially in drier years)."[14] Techniques have now been developed for building vernal pools that overcome the problems associated with the short amount of time artificial pools may hold water reported by these authors.[15]

I spoke with a number of biologists interested in herps (amphibians and reptiles) about the possible benefits of establishing groups of small wetlands in a forested area to provide breeding habitat for woodland salamanders. Biologist trainee Abra Parkman and I had the opportunity to design such a project in response to a grant opportunity from the Forest Service aimed at improving habitat for reptiles and amphibians. We reviewed maps and discussed the merits of various locations to see where there would be little conflict with other resources in creating a number of wetlands in one area. After much discussion, an old clear-cut unit was found that would

be worth examining in the northern part of the Cumberland Ranger District within the Daniel Boone National Forest.

Virginia pine and tulip poplar trees had been harvested from the sloping 10-acre unit above Clearfork Creek in Rowan County 12 years ago. Now a dense stand of young hardwoods, blackberry, and green briar bushes tore at our faces as we pushed through the area. Clumps of pole-sized poplar and oaks left standing after the timber sale had grown up to sawtimber-sized trees and were scattered throughout the young forest. Our progress along old skid trails was often halted by tangles of fallen snags and the thorns of multiflora rose bushes. We stopped a number of times to use a soil auger to sample soil texture at the more level places between the larger trees. Finding silt loam soils in each hole, we began searching for gaps in the canopy that were not on steep ground, marking the perimeter of each gap with hot-pink plastic flagging. There was no way we could see to measure slope or to set a benchmark to indicate the elevation of each dam in the thick vegetation. Our next job was to find a way for equipment to access each area by going around the larger trees, and to avoid confusion with the already-marked wetlands, we used orange flagging to designate these pathways.

Our working in the old clear-cut provided several advantages. There was little risk that trees suitable for use by the federally endangered Indiana bat would be removed since all the merchantable trees had been cut. There were no concerns from foresters that the wetlands would impede future harvest operations, because, again, all the trees had been cut. Access roads were already in place, and since the cut stumps had rotted, they no longer presented an obstacle to construction.

Abra and I marked ten small wetlands ranging in size from 0.2 to 0.5 acre for construction in the harvest unit. Some of the wetlands were only 100 feet from each other. Slope within the harvest area averaged 3 percent, so we selected individual wetland sites that appeared to have less than a 3-foot change in elevation from their upper to lower edges. We planned for the wetlands to be from 12 to 18 inches deep, so that they would hold water long enough for salamander larvae to develop. Seven would be built shallow enough to dry annually, and three would be dug deeper so they would only dry in drought years. Each site had a small watershed, averaging less than two times the area being flooded.

While Abra returned to the thorny regeneration area to prepare a contract map with GPS equipment, I prepared a biological evaluation to document potential effects of implementing the project on endangered, threatened, and Forest Service sensitive species. We also requested a cultural resource survey from District Archaeologist Frank Bodkin. Frank assigned Richard Hunter and George Morrison, trained forestry and cultural resource technicians, the job of field testing the harvest unit. They did not find signs of historic or prehistoric occupation. Frank then wrote a cultural resource survey report on the proposed project for review and approval by the State Historic Preservation Office.

A little more than 6 months later, District Ranger Dave Manner signed a decision memo for the project, which also included a number of other wetlands projects to be completed within the Cumberland Ranger District. The decision memo satisfied the National Environmental Policy Act requirement for documenting projects that affect federal lands.

I developed the following specifications for a contract that was advertised and awarded in 2004 for building the wetlands in the timber harvest area:

Work involves constructing 10 shallow water vernal pond wetlands on the Daniel Boone National Forest, Cumberland Ranger District in Rowan County, Kentucky near Clearfork Creek as shown on the attached map [not attached here].

Equipment required for the job will be one dozer equivalent in weight and horsepower to a Case 1150E. The designated Agency Representative will determine when conditions are suitable for work and will direct the contract.

The work sites are located in an area clear-cut about 12 years ago that is now growing up to dense sapling and pole size trees. The wetlands are basically being built in the lower stocked, shrubby areas between larger trees. Work sites range from 0.2 to 0.5 acres, averaging 0.3 acres in size. On the average, the dam on each wetland will be about 150 feet long and 20 inches high. Wetlands will be made to contain water no deeper than 18 inches.

Required actions for building each vernal pond wetland:

1. Remove trees and shrubs from each work site with a dozer as marked on the ground by plastic flags. Roll the trees cleared with the dozer to remove excess topsoil from the roots. Pile cleared trees, shrubs and debris at a location adjacent to the work sites as designated by the Agency Representative. From two to three larger diameter trees will be saved where available to be later placed in each constructed wetland.

2. Remove and save approximately one-third of the topsoil at each wetland site, stockpiling the topsoil at locations designated by the Agency Representative (generally along the upper edge of each wetland). The topsoil will be later spread in the bottom of each constructed wetland by the contractor.

3. Remove approximately two-thirds of the topsoil remaining at each work site, piling this topsoil in front of the flags along the lower edge of the area cleared so that it can be used to build the backside of the dam.

4. Use a dozer to make a core beneath the location of each dam being built. The center of each dam and the location of the core will be designated by wire flags placed by the Agency Representative. The core and center of the dam will generally be located one-third of the way uphill from the lower edge of the cleared area so there is room for the dam and back slope without crowding the adjacent trees. The core trench will be the width of a dozer blade, approximately 12 feet wide, and extend down to bedrock or an impermeable layer of soil as determined by the Agency Representative. The core will go down to the bottom of any crayfish burrows that are present and will cut through all tree roots. Soil

suitable for the dam and core does not have to be removed from the core, but does need to be shifted to collapse holes, and then packed with the dozer. Small patches of rock, roots, or logs may need to be removed from the core so that water will not travel under the dam. Soil removed from the core trench that is not suitable for placement in the core will be placed outside of the wetland, and can be used to make the backside of the dam. The Agency Representative will make the final determination of which soil is suitable for use in the core and dam.

5. Additional soil needed to build the core and dam will be taken from inside the designated pool area of each wetland. Soil will be removed starting from the uphill, back edge of each cleared area and proceed over the future pool area to maximize the surface area of water in each wetland. Soil placed in the core will be packed with a dozer with lifts no greater than 6 inches thick. The soil in the core will be packed by running over each layer with a dozer the number of times determined by the Agency Representative until suitable compaction is obtained.

6. Build a dam above the core with suitable soils, using lifts no greater than 6 inches thick, packing each lift with a dozer. The top of the dam will be the width of one dozer blade, approximately 12 feet wide. The inside slope of the dam will be a 10:1 slope or more gradual, the backside of the dam will be a 5:1 slope or more gradual. The dam will be built to the elevation of the benchmark designated at each site, generally a nail placed in a nearby tree by the Agency Representative. The top of the dam will be made level (± 0.1 feet) with the elevation of the benchmark.

7. A spillway will be constructed for each wetland to an elevation of 0.5 feet below the top of the dam at one end of the dam so that water will flow out of the wetland around the dam over undisturbed soil. The final decision on the location for the spillway will be made by the Agency Representative.

8. Spread the saved topsoil over the bottom of the finished wetland and up on the inside slope of the dam when finished. Also place larger trees

that were saved in the bottom and around each wetland for use by wildlife and to help block ATV access.

We received six bids for the project that ranged from $4,524 to $7,648. The contract was awarded to John Utterback, a qualified bidder who submitted the lowest price.

John Utterback and his uncle Anthony Utterback moved in two dozers to complete the project in two long days during the last week of July 2004. Forest Service Technician Richard Hunter and I supervised the contract, with one of us being on-site the entire time they were working. The Utterbacks had never built ponds that were this shallow before, so we had to reassure them that we indeed wanted them to be shallow enough to dry out by fall. We made one brush pile from clearing each work site and tried to hide the debris behind clumps of larger trees. As each site was cleared of vegetation, we used a tripod-mounted level and rod to measure slope. We calculated the height of each dam to be the difference in elevation from the upper to the lower edge of each clearing, adding 6 inches for the spillway. We drove a nail in a tree along the lower edge of each clearing at the same elevation as the dam to serve as a guide during construction.

To help make the shape of two of the wetlands appear more natural, we identified the need to clear additional ground during construction. The change to the project was negotiated and agreed upon with John Utterback at the time, adding $400 to the contract. I have learned to expect modifications to wetland contracts while they are under way, and I routinely set aside from 10 to 15 percent of the contract cost for these changes. What will be found beneath the surface cannot always be anticipated, and contractors are more likely to keep their bid low if they know you are prepared to pay them for unexpected problems.

George Morrison and Abra placed branches in the new wetlands to serve as egg-attachment sites for salamanders and perches for birds soon after they were constructed. They also seeded the exposed soils above the water level of each wetland

Construction locations marked by orange-colored plastic flagging for wetlands near Clearfork Creek in a 12-year-old timber harvest area.

John Utterback using an 1150E Case dozer to remove trees and shrubs from the construction site.

About two-thirds of the topsoil is removed and piled behind the dam location.

Dozer cuts a core on top of the dam location.

Core cuts through tree roots and is free of logs and gravel.

Core is filled and packed with silt loam soil.

Dozer backs up to upper edge of the cleared area to begin moving soil for the dam, thereby maximizing surface area instead of depth.

Dam is built in layers packed on top of the core.

A gradual slope is placed on the dam.

Height of the dam and depth of the wetland are measured by rod and level before final grading is done.

Topsoil is shaped into the back slope of the dam, and the top of the dam as well as the rest of the dam is packed and leveled.

Spillway is positioned at one end of the dam, 6 inches lower than the top of the dam, to carry overflow around the dam on undisturbed soil.

Gentle slope is placed on the uphill edge of the wetland.

Soils exposed above water are seeded to winter wheat and mulched with straw to prevent erosion.

Ephemeral wetland the first winter after construction.

with a mixture of wheat, endophyte-free fescue, and ladino clover, then mulched exposed soils with straw to prevent erosion.

The ten wetlands we built filled with water within 2 weeks of construction and stayed full the rest of the year with all the heavy rains brought in by hurricane-related storms.

After building a number of small wetlands in timber harvest units, I have found a common problem that contractors create for themselves during construction. All too often, they push soil for the dam too close to the edge of the area they just cleared for the wetland. This creates difficulty later as no room remains for packing the backside of the dam or for placing a gradual backslope on the dam. To prevent this, I tell the contractor to expect that the wetland they are building will only flood about one-half of the area they have cleared, and that they should then windrow the soil to be used for the backside of the dam from 20 to 30 feet away from the entire lower edge of the clearing.

Mountaintop Wetlands

Mountaintop wetlands occur naturally on the Cumberland Plateau within the Daniel Boone National Forest in eastern Kentucky. They tend to be small, averaging 0.25 acre in size, and can be located in shallow clay and silt loam depressions along narrow ridgetops averaging 1,100 feet in elevation. These mountaintop wetlands range from emergent, ephemeral, scrub-shrub to wet meadow types and are generally isolated from other streams and wetlands in a watershed. Many of these uncommon wetlands were drained for agriculture before becoming National Forest System land. Others were drained for roads, and some were even deepened for watering livestock and white-tailed deer.

I began working in the Daniel Boone National Forest to create and restore mountaintop wetlands in 1988. Apparently, I wasn't the first to do so, as we occasionally find deepwater ponds constructed on ridgetops from the 1950s. Since 1988,

I have worked with technicians and biologists to build over 800 ridgetop wetlands in the Daniel Boone National Forest to improve habitat for a variety of wildlife species. We generally construct these mountaintop wetlands within the oak-hickory forest on clay and silt loam soils, within small dry depressions, and on slopes ranging to 6 percent. These wetlands have been built with water depths ranging from 6 to 60 inches, with an average diameter of 40 feet.

To find locations suitable for wetland construction, I walk mountain ridges and look for shallow depressions or level areas between larger trees. I like to find an area with a diameter of around 70 feet between large trees so there is enough room for heavy equipment to operate. I then use a soil auger to check for silt loam or clay soils. Soil maps do not have the detail necessary for working on this small scale, and even when they show rocky or sandy soils, I've often been able to find clay and silt inclusions that are suitable for construction. I then mark the perimeter of the site to be cleared of vegetation with bright-colored plastic flagging, and I proceed to mark an access route to the site through the woods with different-colored ribbons.

Forestry Technician Richard Hunter found looking for white oak trees to be a more efficient way to identify sites where soils are suitable for building wetlands in the mountains of eastern Kentucky. For years he and I have walked ridges in the Daniel Boone National Forest to mark locations for wetland construction. I always carry a soil auger, while he uses his eyes. Upon finding a somewhat level area in a gap between large trees, he will watch patiently while I use the auger to sample soils. After uncovering soils of high clay content, Richard can always point to a white oak tree nearby that he used as a clue to the presence of fine-textured soils.

I have found that by using a small dozer like a John Deere 650, a road does not have to be built into the work site. This helps reduce problems associated with road building, such as erosion and inadvertently opening up trails to all-terrain vehicles (ATVs). The dozer operator is asked to keep the blade raised on the way in to and out from the site, and to go around trees. This also reduces the amount of area that needs to be surveyed for cultural resources. We typically build two of these ridgetop wetlands a day at an average cost of $600 each, which includes personnel, equipment, and revegetation needs.

Over the years we have found that ten species of bats make use of constructed ridgetop wetlands, including the federally endangered Indiana bat and the Virginia big-eared bat, along with the Forest Service sensitive Rafinesque's big-eared bat and eastern small-footed bat. These small wetlands have been found to be excellent places to capture rare bats during the summer months. White-tailed deer, black bear, and wild turkey are game species that the public commonly sees using the wetlands. We have found that the American toad, Fowler's toad, marbled salamander, spotted salamander, Jefferson's salamander, and wood frog breed in the constructed ephemeral wetlands, and the green frog, gray tree frog, leopard frog, pickerel frog, bullfrog, and common newt breed in the permanent water pools. Birds commonly drink from and bathe in the small forested wetlands, and even the Louisiana water thrush, wood duck, and common merganser frequent these isolated habitats.

Borrow Pits

Engineers often remove vast quantities of soil from sites they call "borrow pits" to complete road and dam construction projects. Most attempts to vegetate these disturbed areas have failed because of the small quantities of seed, lime, and fertilizer applied to them after soil extraction. Borrow pits often remain bare and continue to erode for years after use, leaving ugly scars on the land. I have found that borrow pits generally provide excellent sites for wetland establishment and that wetlands constructed from them will provide habitat for a diversity of plant and animal species.

Soils are often removed down to a layer of rock in a borrow pit, and, generally, this rock layer can

(top) Fifteen-year-old emergent wetland on a ridgetop in Rowan County within the Daniel Boone National Forest.

(bottom) Five-year-old ephemeral wetland on a ridgetop in Morgan County within the Daniel Boone National Forest.

serve as an impermeable foundation on which to build a dam. Finding enough soil to construct the dam can pose a problem in some borrow pits, and often the soils that remain are so compacted that it is necessary to use heavy-duty rippers attached to a dozer to loosen the ground for construction. A benefit to using borrow pits can be that the expense of an archaeological survey is seldom re-

quired in consideration of the massive ground disturbance that has already occurred.

Individuals have found innovative ways to build wetlands in forests and fields, on mountaintops and slopes where the soils are dry and the water table is low. Their success comes from the ability to construct basins that reduce how rapidly water soaks into the ground. Each found it necessary to compact fine-textured soils and to excavate deep into the ground along the lower edge of the wetland to make sure that water would not travel beneath the dam via drainpipes, roots, crayfish burrows, and permeable soil layers.

(top) Ephemeral wetland constructed in 1999 by blocking a ditch in a 25-year-old borrow pit used to build the Zilpo Campground within the Daniel Boone National Forest. The wetland failed to hold water as planned until the dam was cored, which collapsed two crayfish burrows that had been left in a ditch filled by the dam during construction.

(bottom) Small, 11-year-old ephemeral wetland established with a dozer by Dick White in a borrow pit with a high water table in Bath County, Kentucky.

CHAPTER THIRTEEN

Building Wetlands on Wet Land

Building wetlands in places where you get your feet wet is rewarding yet has its own set of challenges. Generally, most any area that holds some water can be reshaped to hold more. Soils are wet for two main reasons: they have a high water table, or they are underlain by packed, fine-textured soils that retain surface water. The key to successfully building wetlands in wet areas is selecting the right strategy, either dipping into a high water table or shaping and compacting a basin so that it holds more surface water.

Gravel pits can provide suitable locations for wetland establishment, as many excavations dip down below the water table. Contractors and transportation agencies often welcome wetland establishment as a part of gravel pit reclamation to reduce the high costs associated with filling low areas after extraction. The bare ground found around gravel pit wetlands can also provide habitat for migrating shorebirds, thereby increasing wildlife-viewing opportunities in these disturbed areas.

Sand Lake Gravel Pit Reclamation

Highway 2 traverses a major portion of the Superior National Forest in northeastern Minnesota. The paved road was in rough shape, containing numerous breaks and bumps caused by repeated frost action and the pounding of hundreds of logging trucks. Tourists wanted Lake County to rebuild the road, and this would require an enormous amount of gravel.

County officials approached the Forest Service for permission to use the existing Sand Lake gravel pit, located on National Forest System land. The area was sensitive to the public, as visitors traveled an access route through the gravel pit in order to reach privately owned lake homes and to reach a Forest Service campsite on Sand Lake. Smaller amounts of gravel had been removed from the pit over the years; however, these quantities were slight in comparison to what was needed for this project. Development would end up expanding the 6-acre gravel pit to approximately 22 acres in size.

According to the terms of a cooperative agreement between Lake County and the Forest Service, the county would not have to pay for the gravel; however, it would need to obtain a free Special Use Permit from the Forest Service for the gravel. Geologist Stu Behling had concerns about the county obtaining gravel from the pit because he had seen too many gravel pits opened, closed, and seeded, only to be reopened a short time later. He could envision plans specific to individual gravel pits in the Superior National Forest that would document long-term decisions on how to manage the gravel resource. Stu knew the extent of the gravel and wanted to be sure that the amount of gravel needed by the county to rebuild the highway would exhaust the gravel available in the pit, so that the pit could be closed. This, then, would provide the perfect setting for a gravel pit management plan to be developed and implemented, because there was no threat to the "design investment" by future demands for gravel from the pit. The county would also be allowed to use the pit as a staging area for large boulders and to rework asphalt.

Stu Behling and Barb Leuelling, Forest Service soil scientist, described a vision for the gravel pit that would avoid the bare and unsightly appearance of so many other pits that had been simply sloped and seeded after use. Barb took the lead in working with Aurora District Ranger Dave Miller to form an interdisciplinary team (ID team) involving herself, a recreation forester, an engineer, a landscape architect, and a wildlife biologist (myself) to identify a desired future condition for the site. The county would then be required to implement the steps outlined in the gravel pit plan as a condition for obtaining the Special Use Permit.[1]

The ID team prepared an outline that included saving and spreading topsoil, establishing shallow-water wetlands, seeding a variety of grasses to establish openings, planting trees and shrubs, applying lime and fertilizer, distributing large logs, and even shaping vertical-cut banks for swallow and kingfisher nesting. Landscape architect Mike Schrotz prepared an attractive, large-sized drawing that showed conceptual slopes, placement of boulders, and areas of woody debris distribution, which helped everyone see how the reclaimed gravel pit could look. His drawing proved most helpful in eventually gaining approval to proceed with the plan.

Over the next several months, meetings were held where the ID team presented recommendations for managing the Sand Lake gravel pit to upper-level managers. The proposal was not supported by all, as evidenced by these written comments found in the project file that were made by two Forest Service staff officers: (1) "Could be viewed as holding the gravel hostage for the work we demand"; (2) "It will cost us in county relations, they will collect in other ways"; (3) "Looks like a plan to totally blow relations with the county"; (4) "We do nothing this intensive ourselves." After many discussions and meetings, including public forums, and with the strong support of Ranger Miller, the plan and the Special Use Permit were approved for implementation.

Lake County eventually established four emergent wetlands with gradually sloped edges following gravel extraction in 1986. The wetlands ranged from 0.02 to 1 acre in size, averaging 24 inches deep, being designed to hold water by excavating into the water table. Topsoil was saved and spread over much of the disturbed area, but not in the wetlands. Seeding of grasses and legumes, along with fertilizer and lime application, began in 1987 and was finished in 1989. Shrubs and trees were planted in 1989; delays occurred because of difficulty locating sources for the plants and because drought conditions occurred in 1988. The shrubs planted included mountain ash, elderberry, beaked hazel, pin cherry, highbush cranberry, and chokecherry. Many of the shrubs were also planted around the wetlands.

Looking back on the success of the project, Barb Leuelling said, "I'm especially proud of gaining the provision in the Special Use Permit that stipulated there must be a success rate of 70 percent or more in order for the seeding and planting operations to be considered satisfactory and acceptable, and the fact that they were required to stockpile surface soil layers for re-spreading." As the biologist, I was pleased with the outcome because, for the first time, the Forest Service did not require that ponds be filled after gravel was extracted. It was also quite a milestone back in 1987 gaining approval for these new wetlands. There was a widely held belief that the Superior National Forest had plenty of wetlands already established, so the construction of any more would be a useless endeavor.

Since they were built, two of the wetlands have been found to perform as ephemeral wetlands by drying occasionally; the others hold water year-round. When visited by Barb Leuelling in June 2004, the wetlands contained much frog activity, redwing blackbirds, swallows, dragonflies, cattails, duckweed, water lilies, swamp cinquefoil, and sensitive fern.

Gravel Pit Wetland Management

Most gravel pit wetlands are formed by groundwater, which can produce an environment poor in nutrients. The lack of nutrients in gravel pit wetlands is not a large concern, as, over time, these systems can be expected to accumulate nutrients and increase in productivity. Even though they are nutrient poor, I have found considerable amphibian, reptile, and bird use of gravel pit wetlands, especially where predatory fish are lacking. I have also observed a number of shorebird species making use of gravel pits with their sparse vegetation. Uncommon bog plants can also colonize the more acidic, nutrient-poor waters found in gravel pits.

When working to create wetlands in a gravel pit, look for topsoil that may have been piled along the edge when it was first constructed. If found, ask the contractor to spread a layer over the bottom of each wetland being built. Often trees and shrubs removed during the original construction of the gravel pit have decomposed to the point where they can be used as topsoil. Where topsoil is lacking, you may consider spreading hay in the bottom of completed wetlands, and even adding entire bales to increase nutrients. Placing boulders of various sizes in these wetlands can provide birds with perches and turtles with sunning sites.

Consider having the contractor purposely excavate wetlands of various sizes and depths to increase diversity. Since water tables are subject to change, you may want to make some deeper wetlands that will hold water in a drought year. The contractor may also appreciate the additional gravel obtained by this action.

Require that the contractor taper the edge of each new wetland, making gradual slopes down to the water's edge. I often make these slopes as gentle as 20:1. Gradual slopes will result in greater plant diversity in these locations.

I have tossed thousands of branches and moved hundreds of logs into recently established wetlands in order to improve habitat for reptiles, amphibians, and birds. Michael Kenawell reported that constructed wetlands studied within

(top) One of four shallow-water wetlands built in the Sand Lake gravel pit in the Superior National Forest in 1986. (USDA Forest Service photograph)

(bottom) Barb Leuelling at the same emergent wetland at Sand Lake in 2004, some 18 years after establishment. (USDA Forest Service photograph)

the Daniel Boone National Forest supported avifauna diversity and abundance similar to a natural wetland. He found that the size of constructed wetlands proved to be a significant factor for determining richness and abundance of avifauna, with the largest wetland studied (6.4 acres) containing the highest total species richness and total species abundance. He also found that the age of constructed wetlands studied was not correlated with species richness or with obligate species richness, and he believes that the various depths created within individual wetlands along with the presence of woody debris contributed to overall bird use of these areas.[2]

My experience in working with Special Use Permits has shown that agency-established base rates are often less than the current market price of gravel or fill a contractor obtains from National Forest System land. This means that you can often negotiate with the contractor to incorporate many of the landscaping actions needed for improved wetland function as part of a permit at no additional cost to the government.

Explosives and a High Water Table

Even though using explosives to create small, open-water wetlands may not be politically correct, it can be an effective, low-cost way to safely establish ephemeral or emergent wetlands on sites with a high water table. I have successfully used this technique to create wetlands up to 0.1 acre in size in the Superior National Forest in Minnesota and the Daniel Boone National Forest in Kentucky.

Superior National Forest

Denny FitzPatrick, a wildlife technician in the Laurentian Ranger District of the Superior National Forest, demonstrated genuine passion for helping wildlife. His dedication and interest were contagious as he worked from 1985 to 1990 with Paul Tine' and Jon Hakala, forestry technicians, to blast twenty-eight potholes. Tine' and Hakala were certified blasters who had received training in the use of explosives to help fight wildfires where they created fire lines and small pools where pump operators could draw water during initial attack and mop-up.[3]

Denny's goal was to obtain some kind of open water in every section (square mile) within the Laurentian Ranger District that would "help grouse, deer, frogs, ducks, turtles, bugs, and the world in general." He worked to create oval-shaped ponds from 50 feet wide to 100 feet long. When asked how he selected sites for blasting, he replied:

Aerial photos would tell me if there was any open water in a given section. A section without open water was given priority. I did take reassurance soil tests with a posthole digger to see if they filled with water, and I did look at vegetation and vegetation maps. Bog areas and lowland areas were good choices. I had the best luck when I was sure that some sort of water table existed close to the surface. A good choice was always the *edge* of a bog area, where the tree line sloped down to end at the bog. The Langley site was a good example of that. The center of that bog was often too solidified with vegetation, but the edges were less dense and showed a little standing water. I did use soil maps, as they often told me if the bottom of the pothole would have a clay base to hold the water.

In September 1990, one of the wetlands they established was near the Langley Road in St. Louis County. Jon Hakala remembers the crew using chain saws to cut through the organic material on the surface of the floating bog, so they could then use posthole diggers to make holes for the ammonium nitrate mixed with fuel oil (AN/FO). They dug fourteen holes, from 10 to 12 feet apart, 5 to 6 feet deep, in a staggered pattern and placed a 50-pound bag of AN/FO, wrapped in plastic and covered by woven nylon, in each hole. The AN/FO sacks were about 8 inches in diameter by 3 feet long. Jon obtained the AN/FO at a most reasonable price from a neighboring iron ore mine. A detonation cord was placed in what was called a "redundant loop" for setting off the charges. Delays were placed on a portion of the charges to move soil away from the site to prevent it from falling back into the hole. Jon said, "The wind came up, and we thought this was to our benefit because we did not have a lot of soil come down in the hole."

Jon said:

(top left) Dan Koschak and Gerald Struckel use chain saws to cut through thick vegetation so that posthole diggers can be used to create holes for explosives at the Langley site in the Superior National Forest. Paul Tine' is in the background. (USDA Forest Service photograph)

(left) Blast column at the Langley Road site resulting from the detonation of 700 pounds of ammonium nitrate mixed with fuel oil (AN/FO) in St. Louis County. (USDA Forest Service photograph)

(above) Langley site immediately after blasting in 1990. The hole was 30 feet wide by 80 feet long and 8 feet deep. (USDA Forest Service photograph)

(below) Langley wetland in 2004, some 14 years after blasting. (Paul Tine' photograph)

We used a fair amount of AN/FO for the blast, but all agreed that after the work involved in packing it in, we weren't packing it out. . . . We had a finished product of 80 feet by 30 feet by 8 feet in the deepest part of the hole. We had mud fly 200 feet and rock even further. The blast dumped muck all over the Langley Road, which they shoveled off by hand. The hole began to fill with water immediately, and is at least 6 feet deep today.

Denny placed a couple of waterfowl nest boxes near the new wetland, just as he did with all the others they made.

Denny declared:

The nice thing about blasting was that it didn't have to be right by a road; we could backpack the supplies to the site. When I could get heavy equipment like a dozer or backhoe to the site, I would do so, because then I could construct an island in the center as safe refuge for critters. But heavy equipment was more expensive than blasting, so we couldn't afford to do many of them without extra funding from the Ruffed Grouse Society or Ducks Unlimited.

Denny knows that all the ponds they blasted within the Superior National Forest held water and that they continue to hold water today.

Paul Tine' revisited the Langley pond and another wetland he blasted near Otto Lake for the first time since he helped establish them nearly 20 years ago, to find that they looked natural and were teeming with frogs and aquatic plants.[4]

Jon Hakala continues to use explosives for management in the Superior National Forest. These days he says that he prefers to use an emulsion instead of AN/FO as a blasting agent to establish small wetlands. He would purchase 4-pound packages of emulsion in sausage-shaped containers measuring from 4 to 6 inches in diameter by 24 inches long. He suggests tying three of the sausages together and sinking them in holes dug in a circular pattern to make a round pond or staggered to make an oval pond. Using an emulsion has many advantages over AN/FO, as it is lighter to carry, it does not have to be kept dry, and placement involves little digging.[5]

Daniel Boone National Forest

The Meyers Fork area within the Daniel Boone National Forest was labeled "Mires Fork" on early maps for good reason—farmers and their implements would become hopelessly stuck in the narrow muddy fields surrounded by steep mountains. These bottomland fields, most likely wetlands at one time, were probably drained in the late 1700s by straightening the main creek, channeling smaller streams that flowed off the hills, and filling in depressions. The Army Corps of Engineers purchased the land along Meyers Fork for the construction of the Cave Run Lake Reservoir in the late 1960s, transferring ownership to the Forest Service in the early 1970s. Much of the land acquired along Meyers Fork remained above the summer pool of Cave Run Lake. Some of the ditches that carried water from the mountains into Meyers Fork had plugged since the 1970s, spreading water over the surface of an area now dominated by willow, alder, sycamore, and multiflora rose.

In 1995, blasting was used to establish small ephemeral wetlands in the now grown-up fields. A soil auger helped identify places to blast where soils were saturated and the water table was at the surface in the winter. Soils on the site consisted of 2 to 3 feet of sandy silt loam to sand over 3 to 4 feet of washed creek gravel lying on top of bedrock.

Forest Service Silviculturalist Jeffery Lewis took a genuine interest in the blasting project, suggesting that his friend Mike Karr be put to work during a planned visit. Mike was a certified Forest Service blaster who worked on the Umpqua National Forest in Oregon. He agreed to help, prompting the scheduling of the project to begin in January.

The blasting project resulted in the creation of ten permanent water and ephemeral wetlands from 3 to 5 feet deep, 15 to 30 feet wide, and up to 80 feet long. The entire cost of the project, including personnel and supplies, was under $3,600.

A day after the blasting was completed, a private individual phoned the Forest Service and claimed that we had cracked the foundation on the home

Variety of aquatic plants growing 20 years later in wetland created by Paul Tine' and Denny Fitzpatrick in 1984, by blasting near Otto Lake in the Superior National Forest in St. Louis County. (Paul Tine' photograph)

he was building one-third of a mile away across a portion of Cave Run Lake. Forest Service engineering inspection found that the blast had nothing to do with the foundation cracking and that it was most likely caused by unpacked soils settling on the home site. Nonetheless, concern over the possibility of other private individuals making claims against the Forest Service for future blasting projects has kept us from using the technique to establish any more wetlands in the area.

Over the years, we have found that several of the wetlands we blasted dry every fall, while others dry only in drought years. The water level in each wetland varies with the elevation of the water table, which can fluctuate as much as 5 feet a year. Their ephemeral nature is of great value to amphibians, as each pool can be inundated by Cave Run Lake on an irregular basis, which, when it recedes, inadvertently stocks them with predatory sunfish, bass, and muskellunge.

Central Oregon

An interagency team including members of the U.S. Department of the Interior (USDI) Bureau of Reclamation, U.S. Department of Agriculture Forest Service, USDI Geological Survey, USDI Fish and Wildlife Service, Oregon Department of Fish and Wildlife, and Sunriver Nature Center recently used blasting to create an open-water habitat for the rare Oregon spotted frog in central Oregon. At one time, the Oregon spotted frog could be found from southern British Columbia to northeastern California. Populations are now restricted to only one-third of their original range. Hydrological alteration and direct loss of wetlands associated with agriculture and urban development, combined with the stocking of predatory fish, are believed to be major contributors to the species decline.[6]

The impetus for creating wetlands for the

Eight-year-old ephemeral wetland constructed with one 50-pound bag of ammonium nitrate mixed with fuel oil (AN/FO) near Meyers Fork in the Daniel Boone National Forest.

spotted frog came when the Bureau of Reclamation was required to buttress the 2.6-mile-long dam on Wickiup Reservoir. The dam improvements required the replacement of a toe-drain ditch below the dam with a 4-foot-diameter pipe. The open ditch provided breeding and adult habitat for thirty-five to forty-five Oregon spotted frogs. The frogs would need to be relocated before the project began. Christopher Pearl, an ecologist with the U.S. Geological Survey, began working with project partners to locate a suitable new home for this small, yet important, population of frogs.[7]

A boggy meadow complex was located within the Deschutes National Forest near the dam site. The 12.4-acre site contained hydrological characteristics and wetland plants similar to other locations where Oregon spotted frogs are found. Using a soil probe, the team determined that the water table was at or near the surface. However,

with no open water in the boggy meadow, the site was not suitable for spotted frog breeding.

The interagency team first considered using an excavator to dig a series of small ponds in the wet meadow. That option was eventually rejected due to concern that the excavator might find itself trapped in areas of saturated ground. Another concern involved having to reopen closed Forest Service roads for the excavator to access the site. The use of explosives appeared to be the best way to make the open water areas needed by the frogs.

Following the decision to use blasting, the Bureau of Reclamation prepared a contract solicitation titled "Blasting for Frog Relocation." That report described the work site as

heavily saturated with water, comprised primarily of groundwater and sub-surface seeps which flow through the wetland and backed up by a low dam which extends across the lower end of the meadow. A vegetative mass overlay what appears to be a shallow braided channel system, with several pockets of deep water present along the main flow line below the vegetative surface. Water depth in the existing main flow channels is as much as four to six feet over the summer months. Several feet of peat covers a stratified substrate, comprised of a silt and clay layer.[8]

Elsewhere in the contract the peat layer was estimated at 8 to 10 feet deep. Before blasting, the ponds were outlined with stakes and flags at four main locations along the deepest part of the subsurface flow channel.

Gerry Dilley, owner of the Superior Blasting Company of Nampa, Idaho, was awarded the blasting contract. He was excited to tell me why he was chosen for the project, saying that his company's chance to help the frogs was "a special thing" and that "we wanted to leave the place better than what we found." He told me, "We even built sleds to carry materials into the site since motor vehicles were not allowed within 200 feet." It was December by the time they were ready to implement the project, and everyone felt a sense of urgency to get the ponds ready for frogs that spring.[9]

Semigelatin dynamite was used as the blasting agent to create the wetlands. Dilley explained the two main reasons for choosing semigelatin dynamite. First, "the high velocity gelatin does not leave waste behind like AN/FO. When AN/FO (which contains ammonium nitrate and diesel fuel) explodes, it leaves behind petroleum and ammonia, possibly affecting water quality in a new wetland." Gerry's crew tested ammonia levels on the site before and after blasting and found that nitrate concentrations actually fell after the project. The second main advantage of semigelatin over AN/FO is that it works when wet. He said, "We even used electric detonators which are biodegradable to reduce the amount of plastic that usually lies around after a blast."

Gerry used 4-inch soil punches to prepare holes from 4 to 6 feet deep for the blasting agent. They first used an auger to drill through the ice

The endangered Oregon spotted frog. (William P. Leonard photograph)

that covered the site. A pipe, known as a casing, was placed in the newly punched hole to keep it from collapsing. Bags of explosives (3.5 × 24 inches) were placed in the holes and pushed to the bottom with powder poles. The casings were then removed for use in subsequent holes. The tops of the explosive bags were set about 24 inches below the surface. The holes were spaced from 6 to 8 feet apart in grid patterns to create ponds in lines to form channels. Biologists had determined the shape of each pond. The site where they blasted was very wet, with the water table within 1 foot of the surface. The area "almost looked like tundra," said Gerry.

The crew arranged the charges to throw soil as far away from the site as possible, using time lags to keep it from landing back in the hole. The delays detonated the outside perimeter charges just before those on the interior. This type of blast is often used to create more-gradual slopes around pond edges and to reduce the ridge of side-cast soil around the pond perimeter. Two blasts were required on the larger ponds to obtain the desired depth.

The project was located more than 8,000 feet from the closest dwelling. Gerry likes to keep at least 2,000 feet away from buildings and structures when conducting a blast for wetlands.

Maintaining this distance helps save the money it would take to complete pre- and post-blasting surveys of local constructed facilities.

The blasting resulted in the establishment of six ponds ranging in size from 0.06 to 0.3 acres and from 2 to 6 feet deep. Water filled them within 24 hours of detonation. This was not an inexpensive project, as the contract cost alone was $85,000, averaging $14,000 per breeding pond.

Christopher Pearl and members of the interagency team relocated Oregon spotted frog egg masses from the ditch to the new ponds in the spring of 2001. The first summer's monitoring found young spotted frogs near the sites where the egg masses had been placed in the new ponds. Research conducted in the new ponds during 2003 was also encouraging, as the number of Oregon spotted frog egg masses was three times that ever seen in their original drainage ditch habitat. Frog larvae also matured at a larger size, and it was observed that the adults grew faster than they had in the ditch.[10]

John Faber, who served as the project construction inspector for the U.S. Bureau of Reclamation, proposed a theory as to why the frogs were doing better in the wetlands than in the ditch. He said, "That 60-year old ditch carried a tremendous volume of water on a regular basis, and the water would stay cold all summer long since it came from the bottom of the reservoir; those new ponds must contain warmer water that's helping the frogs."[11]

When Christopher was asked to describe any problems he experienced in using blasting to create the wetlands, he answered that the technique worked best on the smaller sites and not as well on the larger wetlands. The smaller ponds (approximately 12 × 25 feet) have remained open and are up to 6 feet deep. In the larger wetlands (roughly 30 × 80 feet), some of the blasted material landed back in the pond. Some organic material from under and around the blasted ponds is returning in portions of the larger ponds. Sedges and rushes can become dense on these new organic substrates and are a threat to open water habitat needed by the frogs for breeding. Don-

ning wetsuits and arming themselves with saws, shovels, and picks, team members have removed portions of the invading vegetation mats in an effort to maintain areas of more open water.

Dozers and a High Water Table

A skilled dozer operator can successfully build a wetland on a location with a high water table. In 1985, Denny FitzPatrick used a dozer with extra-wide pads to build five small wetlands within the Superior National Forest on sites with a high water table. He found that "bogs were often too soft a surface for a heavy dozer to operate, so the best dozer site was a lowland area with a fairly firm surface, but with a known water table to tap into." He found that the technique worked well in gravel pits, as well as near seasonal creeks where the soil maps showed a cobble-rock or clay base that would support the dozer.[12]

Denny constructed the ponds to be around 100 × 100 feet in size, 6 to 8 feet deep, with sloping sides and a 10 × 10-foot island in the middle. He contracted the dozer work, with it taking an average of 6 hours to complete each pond at an average cost of $1,000 per site. I asked Denny if he was present during construction. His reply: "Absolutely, I gave oversight, re-hung flagging, and refined the specs and dimensions as we went." He also seeded grass and clover on the exposed banks immediately afterward to control erosion.

Denny was very happy with these completed dozer-dug wetlands. After visiting one 9 years later, he stated, "It looked very impressive and natural, still holding water well, full of frogs, and signs of waterfowl and deer use."

I have also used a dozer to create small wetlands on high-water-table sites; my first location was near Beaver Creek in the Daniel Boone National Forest. In the summer of 1992, Robert Caudill was building a series of wetlands from silt loam soil that were designed to hold surface runoff. After completing the dam for one of these wetlands, he asked why I was not building another one in the small field between the new wetland and the creek. I explained that the site was too sandy to hold water. He chided me by saying, "You don't have to be a biologist to know a good place to build a pond," and claimed he could successfully install a pond on the site. He added that if he couldn't make it hold water, the Forest Service didn't have to pay him. I challenged him to proceed and stood back to watch.[13]

Robert moved the dozer quickly across the area, skillfully pushing only a small amount of soil to avoid getting stuck. Water boiled to the surface after each pass, covering the dozer tracks in seconds. He didn't worry about compaction, being concerned only with making a shallow depression. He completed the 60-foot-diameter wetland in only an hour, and it was full by the end of the day. The Forest Service paid him $55 for the work, and the wetland still contains water 13 years later. I asked Robert how he knew the wetland would work. His reply: "When I saw that water was pumping out of the ground from the crayfish holes behind the dam we just built, I knew all I had to do was to dig a hole down into the water for it to work." I've met few dozer operators capable of building a wetland on a site we call quicksand. Most who try become hopelessly stuck in only a few minutes.

We built another small wetland with a dozer on a site with a high water table near Carrington Lake in Bath County, Kentucky, in 1991. Dick White, the dozer operator, had completed the dam on a

(top right) One of six open water wetlands created by blasting in the Deschutes National Forest in Oregon; photograph taken the first year after blasting. (Photograph by Christopher Pearl, U.S. Geological Survey)

(middle right) Wetland created to provide breeding habitat for the rare Oregon spotted frog in the Deschutes National Forest; photograph taken in 2004, the fourth spring after blasting. (Photograph by Christopher Pearl, U.S. Geological Survey).

(bottom right) An interagency team in a location where they plan to establish additional wetlands by blasting in the Deschutes National Forest. (Photograph by Christopher Pearl, U.S. Geological Survey)

Wetland dug in 1985 with a dozer on a site with a high water table within the Superior National Forest in Minnesota; photograph taken 19 years later. (Paul Tine' photograph)

larger emergent wetland when we still had a few hours left in the workday. With rain predicted, I didn't want to begin a new larger wetland, so I asked Dick to dig a test hole between the wetland we had just completed and the creek. In the hole, we found 1 foot of silt loam, on top of 3 feet of gravel, over bedrock. I'd typically walk away from such a place but could not help staring at the water as it gushed into the hole from the surrounding gravel layer. I asked Dick to proceed in excavating a depression 2.5 feet deep in the middle that would be around 50 feet in diameter with sloped sides. The firm layer of gravel on top of the bedrock provided the machine with the support needed to keep from getting stuck. By the end of the day, we had created a shallow, natural-appearing depression that was gaining water.[14] Over the years, I've noticed that the wetland contains clear water, a variety of aquatic plants, and is ephemeral in nature by drying in drought years.

Excavators and a High Water Table

There are many advantages to using an excavator to construct wetlands on locations with a high wa-

ter table. They are fast, maneuverable, and inexpensive. It is possible to create a 40-foot-diameter wetland in less than an hour, and since most operators charge around $90/hour, the cost can be minimal. Excavators are difficult to get stuck, meaning you can work them in places a dozer operator would fear to tread. I find that it is most efficient to construct a number of small wetlands of various sizes and depths in one location when an excavator is available.

The Forest Service constructed a series of ephemeral and deeper emergent wetlands in two fields grown up to trees in 2002 along Beaver Creek in Menifee County by using an excavator. To design the project, I walked around the old fields in the winter to identify gaps in the tree canopy where soils were saturated. Using a soil auger to find places where the water table was near the surface, with plastic ribbon I marked sites that averaged 60 feet in diameter. The following summer, we hired contractor Earl J. Osborne for the job at an hourly rate to use a Cat 215 excavator with a boom that could reach 20 feet and a 1150E Case dozer.

Richard Hunter and I guided operator David Cade through the dense vegetation to each work site and requested that he dig circular and oval-shaped ponds from 2 to 4 feet deep. He used the large excavator bucket to draw soil toward him and then quickly rotated the machine to spread the soil outside the flagged area. He moved the excavator around the edge of each marked area while continually digging in from the center until each wetland was complete. He smoothed the soils with the bucket so that no artificial-looking ridge of dirt remained around the wetland. David had a tendency to leave sharp, vertical banks along the outside edge of the wetlands, so I had to stop him several times during the day to ask that he give them a more gradual slope. In several locations, we had trouble finding space to spread the soil, as larger trees grew close to the ribbons. In these places, we used the dozer to push, shape, and spread the soils over a greater distance from the wetland.[15]

We achieved the most natural-appearing results in more open areas that were dominated by shrubs with scattered large trees that we could work around. By the end of a 10-hour day we had excavated six wetlands that were from 30 to 40 feet in diameter. Even though the water table was low at the time of year we did the work, water still seeped into the bottom of each wetland within hours of construction. The elevation of water in each wetland follows the water table by being highest from late fall to summer, and lowest from late summer to early fall.

Reclamation of Farm Ponds

You have to take a quick look over the hill and be careful not to bang your head on the glass to catch a glimpse of a particular dry farm pond while bouncing down the Meyers Fork Road in a pickup truck. The site had been farmed for hundreds of years before becoming part of the Daniel Boone National Forest in the 1970s. I had driven by the 40-year-old pond at the foot of the mountain many times over the years and had never seen more than a couple of inches of water in it. I was determined to walk the dam and the bottom of the pond to see if it could be repaired. Getting to the pond proved to be a challenge. Its dam was guarded by a dense growth of multiflora rose full of thorns, but I considered my goal worth defying their barbs. Working my way through the brush, I found only mud in the bottom of the pond with a couple of clumps of sedge beneath the canopy of trees. I took several soil samples with my auger, only to hit rock a foot down, and found nothing wrong with the dam. I hoped that it might be possible to repair the site.

While we were building wetlands in the area in the summer of 2002, I brought dozer operator Billy Osborne over to the dry pond and asked him to blade a path into the site, and then clear the trees and shrubs from the dam and the area around the dam. After it was cleared, we dug a test hole in the bottom of the pond and found gravel, which looked a lot like what was in the bed of Meyers Fork Creek itself. A little water seeped

into the hole while we contemplated our next move. Since we could not find enough clay soils nearby to repair the core and dam, I decided to write off any hope of getting the site to hold water, so I asked Billy to level the old dam and then grade the site so it could be later mowed with a tractor and bush hog. We made no special effort to ensure the site was level or to compact its soils. To prevent erosion, we simply seeded the exposed soils to winter wheat and went on to the next construction site.

When I returned the next summer, I was surprised to see that we had created a wet meadow wetland, with its saturated soils now growing an impressive variety of aquatic plants. Apparently, the old pond had been built over a spring that emerged from the base of the mountain, just downhill from the bumpy road. The spring supplied the wet meadow with water, and leveling the dry pond probably resulted in restoring the original wetland on the site.

Over the past 10 years, I have observed where developers have filled four constructed farm ponds on private lands in Rowan County, Kentucky. In two of these cases, they have succeeded only in changing the sites from deepwater ponds to wet meadow wetlands. In the other two cases,

Wet meadow wetland restored from an old farm pond, leveled in Menifee County in the Daniel Boone National Forest.

the lands are now dry with no evidence of the ponds' presence. The two wet locations are in the open, receiving full sunlight, and now support a variety of aquatic plants. The soils in the wet meadows are so soft and saturated that their owners have not been able to mow them with a tractor and bush hog for 3 years in a row.

Areas where the water table is high have been commonly used to restore a variety of wetlands ranging from wet meadows, to ephemeral and emergent wetlands, to scrub-shrub swamps. Creating open water on these locations simply involves excavating a depression in the ground that exposes water at the desired depth. The length of time that water remains standing in these wetlands may vary by season and year, since it depends on the elevation of the surrounding water table.

You Can't Plow a Pond

Equipment involved with wetland restoration activities can become hopelessly mired when attempting to move saturated soils. Dozers, scrapers, and backhoes all fail to function in mud. A wetland work site can become a quagmire when a thunderstorm hits during the summer months, as the basins tend to collect water anyway. The occasional blocked drain line that is present on a restoration site can also turn a large area into soft, unpackable mud. It is not possible, nor is it effective, to construct a surface-water wetland from mud. Even if one could pile up enough wet soil to form a dam, such a structure can be expected to shrink, crack, and settle greatly.[16] For these reasons, it is often necessary to further drain a restoration site before its contours can be restored.

Completing further drainage of an impacted wetland appears ludicrous to the ecologically minded. And, to the construction worker, being asked to shape and pack saturated soils sounds just as foolish. In 1888, John Klippart recognized that evaporative losses during the summer months were not enough to dry wetlands in Ohio to the extent necessary for agriculture and that

drainage was essential for removing excess water from the soil.[17]

The need for preconstruction drainage before building a wetland designed to hold surface water is best determined before letting contracts and starting work. It is always stressful, and can be quite costly, to have a bunch of contractors and their heavy equipment standing idle because soils are too soggy to move or pack. The best way to determine if further drainage is needed is to examine the area at the same time of year construction is planned, only one year in advance. One quick way to measure soil moisture is to use a soil auger to test from 1 to 3 feet below the surface. Finding saturated soils or a water table within 3 feet of the surface indicates that the area would be too wet to work with heavy equipment. Open ditches should be dug to drain such locations before construction. These ditches need to be deep enough to temporarily lower the water table and wide enough not to fall in before construction begins. A backhoe can be used to ditch smaller sites, whereas an excavator is more efficient for preparing larger sites. The ditches should be dug a month or more before construction is scheduled, using the same pattern as one would use to drain an area for farming.

The first time I used ditches to drain a site before construction was in 1989. The Forest Service had hired Earl J. Osborne by the hour to provide a Case 1150C dozer that his son Billy would operate to construct five surface-water wetlands near Beaver Creek in Menifee County, Kentucky. The first construction site was an old field that had been farmed since the early 1800s. Evidence of prior drainage activity in the 5-acre field included straightened streams bordering each side of the field, a diversion ditch along the base of the mountain, and a series of shallow waterway ditches that cut through the center of the field.[18]

Earl began constructing the wetland by first stripping topsoil from where the dam would be built and from an area in front of the dam, which would serve as borrow for the dam. He then began building a low dam by pushing soil but got stuck. Richard Hunter, who assisted with the project, remembered that "little volcanoes of water were shooting out of the ground."[19] He was referring to the many crayfish burrows that propelled water each time a dozer passed. The contract stated that Earl would be paid only for productive work, and I regretted to inform him that freeing a buried machine did not meet the criteria for payment.

Anxious to get his dozer freed and to get back to work, Earl drove to the corner store and phoned a wrecker service for help. A huge tow truck showed up 3 hours later, set up on firm ground next to the creek, stretched out a cable that must have been 500 feet long, and pulled the dozer out, charging Earl $75. When Billy returned to work, I noticed he would now only push a small amount of soil with each pass. He told me that if he tried to shove any more, he would get stuck again. Earl told him to stay away from the wet places, but since we were trying to core into saturated soils to block ditches, he was soon stuck again. We tried for hours to get the dozer out by cutting trees and placing the cut logs beneath the dozer tracks, but with no success. Again, Earl made a trip to the store and called the wrecker for help. The same thing happened next day. As we stood next to each other tired, soaked in sweat, and covered in mud, I could see that this wetland was not going to get built, and Earl could see that there was no money in the job, but that is when a solution was born.

Earl informed me that I would need to hire a backhoe to drain the field as he could not push or shape mud. I could see that we needed two dozers on the job, so one could pull the other out when needed. I decided to give John D. Smith a call. We had recently met at a meeting of the Menifee County Fish and Game Club, and I remembered that he ran a backhoe for a living. John sounded interested in what we were trying to do and told me he would come out to the job site the next morning.[20]

When John showed up, he could not help but joke with me as the new wildlife biologist regarding my sudden interest in land drainage. However, I soon realized that I was in good hands because of his strong interest in helping wildlife. After I showed him where we wanted to build the levee, I

New ditches were dug by a backhoe to dry out this site before construction could begin. The restored emergent wetland is located near Beaver Creek on the Daniel Boone National Forest. Great blue heron nests are visible in the background.

was surprised by the locations where he chose to dig the ditches. He dug new ditches in the open grassy parts of the field that were parallel to the old ditches. This saved him the work of trying to clear trees growing in the soft ground along the partially filled ditches and saved me the expense of removing trees we wanted to flood anyway. It was also faster for him to dig in the open, and the ditches were less likely to cave in because they were out of the mud.

In the very wettest areas, John made the ditches wider at the top so they would not cave in. This involved "stepping them back so they don't fluff in," he said. "You may only have to do this for 30 to 40 feet until you get out of the soft ground. This relieves the pressure. A deep, straight-sided ditch in soft ground will cave in, just like releasing a wedge."[21] I watched as he dug ditches through quicksand that were three buckets wide across the top and one bucket deeper in the middle. These ditches would stay open and drain for months.

In one day's time, John dug a long series of ditches in the field that were from 2 to 3 feet deep. We let the area drain over the weekend. I could not believe how much the drainage ditches had helped when we started back to work on Monday. The dozer was now able to push an entire blade full of soil without getting stuck and was even capable of packing the now drier soils placed in the dam.

Within a week, Earl had bought another dozer that his youngest son Donnie began operating at the work site. I observed a psychological advantage to having the second dozer working at the same time, as we all saw how easy it was to get one dozer pulled out of the mud with another dozer. With the completion of the ditching and a second dozer we were able to build five large wetlands near Beaver Creek that summer.

Using construction ditches has made it possible for us to restore natural levees and reshape basins on a number of wetland projects in eastern Kentucky and Ohio. Ditches were needed on these sites because soils were too wet to move, even during the fall, which is the driest time of year. These wetlands had been drained with ditches and buried drain lines over a 100-year period. The lands were then acquired by the Forest Service and not farmed for 10 to 30 years. Over time, some of the ditches partially filled and an occasional buried drain line became plugged. Going into areas where we planned to restore wetlands and draining them further made me feel nervous and guilty at the same time. But this trip over to the dark side always reaped rewards by making it possible to complete wetland restoration projects, and to do so at a reasonable price.

CHAPTER FOURTEEN # Highways and Waterways

Culverts through roads provide great wetland establishment opportunities, and beavers have known this for years. Existing culverts can often be modified to saturate soils upstream, or to pond water, by using the road for a dam. Those steel corrugated culverts that are used in many areas often have only a 20-year life expectancy, so their replacement may offer an opportune time to return wetlands in an area.

I have established small, emergent wetlands by attaching water-control structures to the upstream end of existing culverts. The most successful of these involves using a simple 90-degree elbow, with a vertical section of pipe as an overflow, both being made from PVC pipe joined together with rubber gaskets. These wetlands can generally be built at a low cost since little, if any, earthmoving is required.

Becoming acquainted with the engineer responsible for road construction and maintenance in the area can yield impressive results when working to create or restore wetlands. The engineer can help determine whether or not actions are needed to protect the roadbed from water damage possibly caused by a constructed wetland. In many cases, no additional actions are needed, because the road was constructed of fill that would not be damaged by adjacent waters. In some cases, additional protection can be as simple as building up the road surface with more gravel or as costly as lining the upstream surface of the road bank with a layer of silt loam or clay-textured soil.

PVC SDR (sewer and drain) pipe is an excellent material to use when replacing a culvert that can later be modified to establish a wetland. Fittings with rubber gaskets can be attached to the

pipe; they seal tightly and will not leak. All but the largest-diameter PVC pipes and fittings can be attached and adjusted by hand using an iron bar. Consider attaching the water-control structure to the downstream end of the pipe in areas where beaver are present. Such action will help keep the pipe inlet under water, lowering the chance that beavers will detect and block it with debris. For added insurance against beavers, it is a good idea to attach a 20-foot section of perforated pipe to the inlet that will further diffuse water flow at the upstream end of the pipe. I have also attached water-control structures known as flashboard risers to the upstream end of existing steel corrugated culverts for the purposes of constructing a wetland. I do not recommend this practice as flashboard risers are plagued with problems: they leak, rust like steel culverts, and are difficult to maintain.

Beware of buried drainage structures that can cause wetland failure when using an elevated roadbed as a dam. In 1903, Charles Elliott provided guidance for using clay drain tiles and open ditches to keep roads dry when building over swampy ground. He advocated installing a system of buried tiles to first drain wetlands, and then building a road over the top of the drained wetland and the tile system.[1] It was also common to cross beneath roads with buried pipes when draining a wetland in order to empty water into a deeper outlet.[2] Quincy Ayres and Daniel Scoates in 1928 recommended using open ditches and buried clay tiles to "cut off the water and keep it from reaching the roadbed." They describe how important drainage is to a road when constructing adjacent to wetlands.[3]

Wetland established by modifying a road culvert with PVC pipe in Rowan County in the Daniel Boone National Forest.

I have seen and heard of a number of cases where drain lines were cut through roads years before they were blacktopped, leaving outlets ending in ditches and creeks distant from the original wetlands location. Should a constructed wetland that is built along a road not hold water as planned, it may be necessary to dig a deep trench along the side of the road to locate and block buried drain lines.

Even though it may appear easier to raise the elevation of a culvert to form a wetland adjacent to a road, there are a number of advantages to installing a water-control structure to accomplish the same purpose. Unwanted fish species can be removed from a wetland by opening the water-control structure. Water levels can also be adjusted to create conditions suitable for the establishment of desired aquatic plants. Water-control structures permit the gradual drawing down of water levels to maintain mudflats that are important to resident and migratory shorebirds.

Building Wetlands on Closed Roads

Small wetlands can be built on closed roads to provide habitat for many species of amphibians,

reptiles, mammals, crustaceans, and insects. One study of the Daniel Boone National Forest found that eight species of amphibians used road-rut ponds for breeding. Surface area, depth, and, to a lesser degree, water clarity were important features affecting amphibian selection of these small waters.[4]

Biologists, including myself, have observed bats traveling tree-canopy-covered roads in the Daniel Boone National Forest. We often set mist nets in road corridors and over road-rut ponds to capture bats. Wetlands don't have to be large to be used by bats. During a dry summer, James Kiser, a Forest Service wildlife biologist working in the Wayne National Forest in Ohio, captured eighteen northern long-eared bats and one federally endangered Indiana bat in a mist net he set over a road rut no larger than a dinner plate.[5]

Roads often provide the only locations suitable for wetland establishment in mountainous areas. The extensive excavation necessary to build a road in the mountains generally involves cutting into steep side slopes to produce narrow, level benches that can later be modified to create linear-shaped wetlands. Constructed roads and timber sale skid trails can be effectively closed by building a series of wetlands along their length.

Ruts that hold water in an existing road indicate that wetlands can be made in that road. Road-rut ponds form when water collects in small dips that are repeatedly driven over by vehicles, which pack the fine-textured soils. Even when water-filled ruts are present, it is still a good idea to test soil texture on a site before embarking on a wetland establishment project. Examining soils in exposed cut banks along the edge of a road can serve as a good substitute for digging soil samples from a packed roadbed.

Here are some steps to take when building wetlands in a road:

1. Locate road segments that are fairly level.
2. Examine soils in adjoining cut banks and road shoulders to see if soils high in clay or silt are present.
3. Call in advance to see if utilities such as elec-

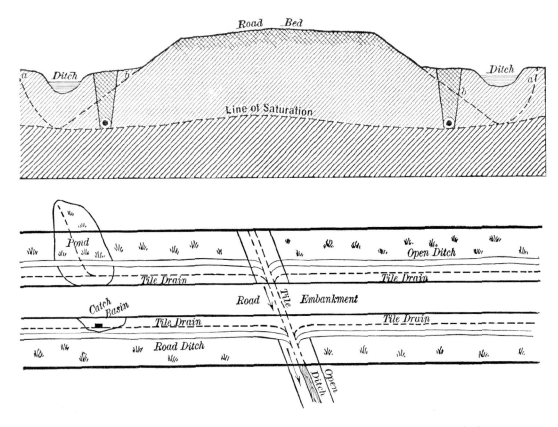

(top) Roadbed drained with ditches and clay tiles. Dashed lines between a and b show how clay tile drainage structures provide the same effect as raising the roadbed to keep it dry. (From Charles G. Elliott, *Practical Farm Drainage* [New York: Wiley, 1903], 88)

(bottom) Sketch of a road constructed over a swampy area and an emergent wetland. (From Charles G. Elliott, *Practical Farm Drainage* [New York: Wiley, 1903], 87)

One in a series of four wetlands built in a row to close an old, entrenched road in the Daniel Boone National Forest in Rowan County, Kentucky.

tric or water lines are buried under the road or along the shoulder.

4. Use a dozer to scrape gravel, woody debris, and branches from the road surface.

5. Use the corner of the dozer blade to excavate a core, taking several passes with the dozer, going deeper each time to make sure that the bottom of the core cuts through topsoil, roots, or loose rock that could carry water under the dam.

6. Place the core down the middle of the road on steep slopes so that enough room remains to build a dam over the core in the roadbed.

7. Fill and pack the core with silt loam or clay soil.

8. Build a low dam with a gradual slope over the core, packing the soil in layers.

9. Make a spillway at one end of the dam to provide a path for excess water to flow around the dam.

Logs and branches removed from the skid road in the timber sale unit within Bath County in the Daniel Boone National Forest. Note the water-filled rut.

Digging a core trench through roots down the middle of the road.

Filling and packing the core with silt loam soil.

Building a low dam with a gradual slope over the core.

Sloping the inside and backside of the dam and packing in layers. (Photograph by Richard Hunter, USDA Forest Service)

Building a spillway to give water a way to flow around the dam.

Small linear wetland completed in 30 minutes.

Small wetland the winter after construction.

Mountain Bogs in the Nantahala National Forest

The restored bog along the Buck Creek Road provides an unexpected glimpse of an attractive ecosystem within the Nantahala National Forest. The mountains of North Carolina once contained a series of bogs, stair-stepped up and down drainages that were maintained by beaver activity. A majority of the historic bogs were converted to farmland in the early 1800s. Land acquisition efforts for the Nantahala National Forest in the upper portions of watersheds have provided opportunities to restore some of these smaller bogs that were drained.

The Buck Creek bog was once much drier than desired for drowning out trees and shrubs. Woody species were taking over the area, shading sphagnum mosses and drying mud used by the endangered bog turtle. Doreen Miller, Forest Service wildlife biologist, had heard that fire should be used to reduce the woody plant invasion. After trying repeatedly to burn these rare communities at different times of the year, she discovered that the mountain bogs within the national forest would not carry a fire. To make matters worse, one of two main streams feeding the Buck Creek bog was cutting a deeper channel down to a bordering road culvert, and this channel was functioning as a drain to further remove water from the bog.[6]

Forest Service engineers were making plans in 1999 to rebuild the Buck Creek Road, so Doreen became involved in the preparation of a biological evaluation designed to document possible effects of the proposed project on federally listed endangered, threatened, and Forest Service sensitive species. The engineering plan she reviewed called for lowering the elevation of the culvert that carried water from the bog. She knew that this action would be the same as digging a drainage ditch in the bog and would reduce habitat available to the bog turtle. With only 20 acres of bogs remaining in the national forest that provided habitat suitable for the bog turtle, such an impact must be avoided.

Hydrology and vegetation in this rare mountain bog were restored by raising the elevation of the culverts during a road reconstruction project within the Nantahala National Forest in North Carolina.

Doreen worked with the district ranger and engineers to modify the project so that the road would actually improve the condition of the bog. They decided to place two large-diameter culverts in the road, setting both at an elevation 1 foot higher than the original culvert. Although there was concern that such a change would result in converting the bog to a large, deep pond, survey work showed that this would not happen.

The area as viewed from the road now looks like a natural beaver pond. A small pool of deep water is found next to the road. This water is bordered by mudflats, and the mudflats are bordered by bog that is basically devoid of trees. Raising the culverts resulted in more acres of saturated soils, and these saturated soils made for a much larger bog. The stream has spread out and stopped down cutting, which has also helped restore natural water movement over the site. Everyone appears pleased with the results, and the bog blends into its surroundings as though it was there when the mountains were formed.

Managers have found that building wetlands near roads can provide unparalleled viewing opportunities. Since many roads were built through wetlands, which often resulted in the wetlands' demise, roadsides are a logical place to complete these projects. Precautions should be taken to protect roads from water damage and to reduce beaver activity that can affect these locations.

Stream Restoration and Wetlands

Arthur C. Parola Jr., professor of civil and environmental engineering and director of the University of Louisville Stream Institute, has developed what can best be called a revolutionary technique in wetland restoration. His method for restoring streams to their more natural condition is also capable of raising the water table over a riparian area, providing an environment whereby wetlands once dependent on groundwater may be restored across the floodplain.

Raising the water table in the floodplain allows for the restoration of wetlands on a scale not feasible by any other practice. Most wetlands are now being restored by using one of two techniques: (1) shaping and packing soils that have high clay content so they will hold surface water, or (2) removing soils to expose an existing high water table. Attempting to build wetlands on sites with permeable soils and a low water table has been problematic, requiring the use of expensive synthetic liners to hold surface waters. However, when the water table is raised by Parola's stream restoration actions, one can simply use heavy equipment to create dips that will fill with groundwater and which are natural in appearance. Once the water table is returned to its historic elevation, low places along the creek may also contain standing water, and the soils in more level areas are likely to become saturated to form wet meadow, scrub-shrub, and forested wetlands. Art's actions are capable of raising the elevation of a water table by 6 feet or more in portions of a floodplain.

Beginning in the 1800s, the majority of streams east of the Mississippi were moved and straightened to create agricultural land. He says, "In most areas I can't find a stream that hasn't been moved." Parola explains that when a stream is straightened, its length is reduced. The shortened stream has a steeper gradient, which causes water to flow at a higher velocity. Over time, the straightened stream will modify its slope by cutting a deeper channel along its length. The vertical sides of the deepened channel will become unstable and, eventually, cave in to form an even wider channel. Straightened streams also function like drainage ditches by lowering the water table and drying adjacent wetlands that were once maintained by groundwater.[7]

Farmers still recognize the value of these deep, straight stream channels for keeping their fields dry, and they work to keep their courses free of fallen trees and logs, which can trap sediments and raise the water table. Historically, many Appalachian streams also served as roads, so that sediments and gravels were kept mobile by animal hooves and the wheels of carts moving up and down them before and after rains. Many of these straightened streams have since eroded down to bedrock and lack the number and diversity of pools and riffles found in unaltered areas.

Small bends can develop in sections of straightened streams that are not kept free of logs and debris. Art has shown me where bows in creeks (up to 15 feet off center) have formed where large trees fell into these channels. The condition is especially apparent in streams that were once farmed by private individuals but are now in public ownership and are no longer being used for crops. Art warns that because of the great depth and width of these straightened channels, simply adding woody debris will not be enough to return them to their original path and depth. He expects that these streams will function as drainage ditches for hundreds, if not thousands, of years without active restoration.

Parola's goal is to actively restore the most natural depth, length, and shape to a creek. Whereas moved creeks basically flow along one side of the valley, he works to make them flow from side to side across the width of the floodplain. Moved stream channels are often deep and wide, so he generally restores them to a shallow, narrow state. His actions reconnect the stream to its floodplain to ensure that floodwaters will regularly flow over its banks and spread across the riparian area a number of times each year, as compared to once in only 100 years previously.

Much figuring goes into his determining the best location, slope, width, and depth of a stream channel being reconstructed. Careful measurements are taken to calculate stream flow under both normal and flood conditions. Buckets are buried in the creek bottom to serve as sediment traps well in advance of restoration to monitor the size and quantity of gravels being moved by heavy rains. Gauges are set along the creek to track the depth of floodwaters, and calculations are used to design a creek that will develop a diversity of riffles and pools that will not be flushed clean of gravel and woody debris.

One of the projects Art Parola, his staff, and I are working on is found in the Daniel Boone National Forest in Kentucky, along Slabcamp and Stonecoal Creeks. Personnel from a number of agencies are planning to restore these creeks with funding from the Army Corps of Engineers Fee in Lieu of Mitigation Program, being administered by the Kentucky Department of Fish and Wildlife Resources.

These creeks are typical of many others in Appalachia where trees had been cleared in their floodplains and sections moved along the base of the mountains to facilitate farming. The Forest Service bought the lands along Slabcamp and Stonecoal Creeks in the 1930s, and the agricultural fields have since grown up to trees. Since that time, several timber sales have taken place near the creeks, and trees were hauled down using the creeks as makeshift roads. A number of creek sections now resemble deep ditches flowing on bedrock. These channels are keeping the water table low, thereby hindering wetlands from forming and preventing waters from flowing over the riparian area.

The straightened streams contain few pools, and the remnant pools are narrow and shallow. Riffles are uncommon, and those present are embedded with fine-grained sediment, which provides poor habitat for aquatic life. Steep, eroding banks border the stream sections moved years ago, clouding the waters with silt and clay sedi-

ment after heavy rains. A portion of Slabcamp Creek has three artificial channels, all with water flowing downhill in the path of abandoned roadbeds. The movement of aquatic organisms up- and downstream, in both the main creek and its tributaries, is now restricted in a number of places, particularly at road crossings.

Undoubtedly, most people who walk along Slabcamp or Stonecoal Creek would find that the streams and forest appear natural, and they may not see the need for restoration. Trees growing along the stream now mask straightened channels that flow along the toe of the hills. Because the water table has been lowered, it is uncommon to find the trees and shrubs that would normally be growing on wet soils along the creeks. Deep pools that support game fish are not present. Smaller streams disappear underground, being carried in culverts, old wooden boxes, rock channels, and clay drain tiles laid in the early 1900s, before the area became part of the Daniel Boone National Forest.

As we walked upstream, Art described how management actions would return a more natural and historic flood pattern to the streams. Waters could be made to disperse and flow gradually over the riparian area in flood events rather than rushing down deep and narrow channels. Whereas the straightened channels now force water against the mountainside, causing much erosion, the restored creeks would meander over the valley floor at a lazy speed, depositing leaves and branches for the benefit of aquatic life.

The downstream reaches of these creeks contain 8-foot-deep channels more than 20 feet wide, running on solid bedrock. The restored creeks in these areas may be less than 10 feet wide with flood-gorged depths less than 2 feet deep, flowing naturally over gravels. In the middle sections, a braided segment of stream channels could be constructed, each possibly less than 5 feet wide and less than 1 foot deep. There may be only a series of small channels, on the order of 1 foot wide, in the upper portions of the watershed. Most of the flow would occur through a zone of saturated soils forming sinuous, wide wet-meadow and shrub wetlands.

A path varying in width from 12 to 60 feet would be cleared for establishing the stream channels. Live and dead trees that are in the designated path of the restored stream would be felled and used to improve in-stream habitat, or removed. Small, naturally appearing wetlands of various sizes and depths would be shaped along the streams. These wetlands, some of which will be directly connected, as well as others that will be separated and connected only during flood events, will play a vital role in filtering sediments and nutrients, reducing flood levels, and providing habitat diversity.

On a beautiful spring day, Art Parola and I accompanied ichthyologist Matthew Thomas, of the Kentucky Department of Fish and Wildlife Resources, on a trip to sample aquatic organisms in Slabcamp and Stonecoal Creeks. Matthew collected fishes by stunning them with an electro-shocker he carried on his back.[8] This battery-powered machine temporarily immobilizes fish so they can be scooped into nets, identified, and then released back into the water. We found that Slabcamp and Stonecoal Creeks provided habitat for the redside dace, an attractive small fish that wears a brilliant red slash on its side. In Kentucky, its distribution is limited to a small number of headwater creeks on the Appalachian Plateau.

Some of the fishes Matthew collected during the day were hybrids of the redside dace, including redside dace/creek chub and redside dace/redbelly dace crosses. The redside dace requires clear, cool water with pebble, gravel, and sand substrates for breeding.[9] "These hybrids generally indicate impaired habitat," he said. "When adequate habitat is not available, more than one species can breed over the limited number of suitable sites, resulting in hybrids that may or may not be fertile."

Parola pointed out how a majority of the creek beds were lined with large rocks and long stretches of bedrock without sand and gravel. "The fines have been washed downstream by fast-flowing waters. Our restoration will change all that," he said. It looks like we are just beginning to see some of the many benefits of stream restoration.

When Art designs a project, he takes steps to prevent restored creeks from reverting into their old, deep channels. Portions of the old channel are dammed with soils high in clay, and woody debris is strategically placed to form a series of stair-stepped wetlands within the previous route.

In some situations, Parola has found it necessary to lower the elevation of the riparian area with scrapers and dozers to remove an accumulation of sediments. The presence of historically deposited soil layers from upstream erosion can be determined by looking at the soil profile exposed by eroded creek banks, or by taking soil borings to identify buried leaf and topsoil layers beneath the surface. Failing to remove these historically deposited layers can make it almost impossible to restore the stream and raise the water table to its previous elevation.

A restored stream must be protected from downstream events that can cause the entire system to unravel. Historic, as well as future, actions that involve straightening, channeling, and gravel removal can seriously affect sections restored upstream. Streams are dynamic and will continue to erode and deepen years after a change was made downstream. A stream can react to these downstream actions by starting a head cut to decrease its slope. A head cut is a vertical drop-off that looks like a miniwaterfall in the creek. Head cuts are not stable and can move fast or very slowly upstream, depending on soil type and the presence or absence of large tree roots that keep a channel from washing. A head cut marks a wave of deepening that can take generations to migrate upstream. Each one is capable of lowering the creek by several feet or more. All the beneficial effects of a stream and wetland restoration project can be lost if a head cut migrates through the restored section of stream. This is the reason it is essential to install grade-control structures in the valley along the lower limits of the creek section being restored that will prevent a head cut from moving upstream into the project area.

There are a number of features that can serve as grade control. These include a series of logs or large rocks buried across the riparian area, a cul-

Natural Stream Channel

Head Cut

Changed and Deepened
Stream Channel

Vertical head cut moving upstream after stream-straightening activity downstream. The 2-foot-deep head cut is migrating upstream on the author's farm at a rate of 4 feet each year.

vert in a road crossing, or even a low-water crossing made of cement. It is important to remember that, should one of these grade-control structures ever be modified, it must be replaced with a structure at the same elevation to prevent a new head cut from forming and moving upstream.

Restoring streams and wetlands in the same place at the same time is logical when you consider historically how farmland was improved by moving streams and filling wetlands simultaneously. Unfortunately, at present, many stream and wetland projects are being conducted separately. Combining stream and wetland restoration into one action at the same place and time is likely to result in greater efficiency when securing permits, conducting surveys, and hiring heavy-equipment contractors. A partnership between those involved in stream restoration and wetland restoration is certain to provide benefits to the countless species of plants and animals that depend on both of these ecosystems.

Do
Waders Come
in Size 4?

CHAPTER FIFTEEN

Many educators are finding that a small constructed wetland makes a valuable and attractive addition to the outdoor classroom at their school. Creating a wetland within easy walking distance of a school provides a ready laboratory for ecological studies. A school's science curriculum is greatly enhanced by involving students in the actual sampling of invertebrates and vertebrates in a wetland. Students can better retain the principles of chemistry by witnessing the reduction equation in action when both seeing and handling wetland soils. Educators have said that more than the teaching of sciences benefits from constructed wetlands, as even art students will use the environments for painting the beauty of a hibiscus or sketching the slender curve of a rush.

In the fall of 2003, science teacher Loretta Roach built an ephemeral wetland at the Dewitt Elementary School in Knox County, Kentucky. She used a $1,500 grant from PRIDE to complete the project. Eastern Kentucky PRIDE is a nonprofit organization that offers environmental grants up to $5,000 to other nonprofit organizations and educational facilities for building and improving outdoor classrooms in a thirty-eight-county area.[1]

I provided technical assistance as Loretta engaged forty seventh- and eighth-grade science students and some of their parents in the successful restoration of a wetland only a baseball's throw from the school. We began early on a beautiful Saturday morning by looking at slides of wetlands and talking about the many plants and animals that depend on them. Everyone ventured outside to look for clues of drained wetlands and got their hands dirty testing soil textures. The group stepped back to watch as the dozer operator began work. I stopped the dozer operator occasionally to explain each stage of construction as it began.

The students helped shoot elevations with the level and determined how large and how deep the wetland was going to be. Excitement grew when the dozer unearthed and blocked two clay tile lines, one of which had been buried in the bottom of a wide, open ditch bisecting the site.

Loretta now regularly takes her middle-school science students outside to examine the wetland:

I talk to them about the need for wetlands and how important they are to the environment. My principal knows to look for us at the wetland if we're not inside the classroom. We'll watch dragonflies hunting and catch tadpoles. Sometimes we take water samples back to the classroom and look under the microscope. When we're out there, they don't even realize they're learning.

And the students' reaction to the project?

The kids are so proud of the wetland—they were thrilled to be involved. The eighth graders who have moved up keep coming back to see it. They can be very protective of the wetland. The other day, a student

182

alerted me that our custodian was getting close to the Indian grass we planted with the mower.

Loretta explained why she worked so hard to build the wetland: "I want to make a lasting impression on the kids. The wetland has been a tremendous teaching tool for them."[2]

Creating a wetland large enough for an entire class to circle provides the best opportunity for everyone to share in the excitement associated with sampling these aquatic environments. A wetland constructed with a minimum diameter of 30 feet and a circumference of 94 feet provides each student with 3 feet of space when a class of thirty circles a pond. Constructing an ephemeral wetland no more than 12 inches deep can help respond to safety concerns. Safety concerns with standing water are generally based on swimming pools and bathtubs, which have hard surfaces. Because wetlands have soft soils in the bottom, someone who falls in is much less likely to be injured on the yielding surface. Even a shallow wetland will contain a wide variety of amphibians, aquatic insects, crustaceans, and aquatic plants suitable for study.

Wetlands add beauty to a school's grounds. Their shimmering waters framed by green rushes look superior to most flower beds. They are always a topic for discussion when the unexpected deer, wood duck, or great blue heron stops by for a visit. Administrators find that wetlands require much less maintenance than a bird bath or fountain.

Before constructing a wetland at a school, it is wise to inform people who live near the schoolgrounds and parents. Be prepared to answer questions about mosquitoes, safety, and the appearance of the wetland. The principal and superintendent should be well aware of what is planned and be prepared to voice their support for the project.

One rarely finds soils that are suitable for building a wetland near a school, so most often a synthetic liner is needed to obtain the hydrologic regime necessary for producing aquatic soils and plants. Extensive earthmoving is done to cre-ate the large level areas necessary for the school building, parking lot, and playground. Cement, asphalt, trees, and woody debris are often buried to create these high and well-drained areas. Finding a little gravel, concrete, or charcoal near the surface is like the tip of the iceberg that indicates much more debris is buried below.

Small wetlands can be built on sites that have a thin layer of soil over buried fill or on slopes where neighboring soil cannot be moved to build the dam by hauling in additional soil for the project. It generally requires three tandem-axle dump truck loads of rock-free soil to create a small wetland that is 30 feet in diameter and 12 inches deep, on a location where a depression cannot be excavated. The first load of soil is used to level the location, the second to form a depression by making a gradually sloped dam, and the third to cover the synthetic liner and help blend the edges of the wetland into the surrounding landscape.

School grounds are typically laced with buried drain lines and gas, electric, water, and sewer lines. It is not only wise but an essential investment of time to thoroughly investigate and determine if buried utilities are present before disturbing soils near a school.

Creating an outdoor classroom with a wetland can radically change and improve appearances of the typically mowed and manicured school grounds. To help lessen the visual impact and gain a greater acceptance of a wetland, consider adding features to your project that landscape architects call "cues to care." A simple split rail or board fence around the outdoor classroom can be a cue to care showing visitors, administration, and especially the groundskeeper that the area behind the fence is something special.

Phyllis Allison proves that you can build a natural-appearing wetland in an urban setting. She has taught kindergarten at the Prestonsburg Elementary School in downtown Prestonsburg, Kentucky, for 20 years. In 2000, she took the lead in building a small ephemeral wetland on their school grounds with financial help from the PRIDE program. "We have a large amount of students who come from an urban setting and

Phyllis Allison and schoolchildren with the ephemeral wetland they constructed.

housing projects that would not normally see a wetland. This gives them an opportunity to share in the outdoors," she says. The group used a small dozer to excavate the depression, then buried a synthetic liner so it would hold water. The attractive array of aquatic plants now growing in the wetland was obtained from donor populations such as roadside ditches. "Our students love to see the dragonflies and tadpoles."[3]

Healthy Wetlands Devour Mosquitoes

Propose a wetland project near a home or school, and you'll soon be asked questions about mosquitoes. People have arduously filled, drained, and sprayed wetlands to control mosquitoes for generations, so it is important that you have an understanding of wetlands and mosquitoes and design your project so it doesn't contribute to the problem.

For over 20 years I have made a special point of looking for mosquitoes and their larvae while visiting natural and constructed wetlands across North America. Mosquito larvae, known as wrigglers, are easily seen in the water and can be readily captured in fine-mesh nets. I've observed that mosquito larvae are absent in healthy wetlands. A healthy wetland contains clean water and therefore provides habitat for mosquito predators such as salamander larvae, newts, adult frogs and toads, dragonflies, damselflies, water boatmen, water striders, whirligig beetles, backswimmers, giant water bugs, and predacious diving beetles. You've probably seen ducks, shorebirds, tree swallows, and purple martins feeding at healthy wetlands during the day and bats at night. Healthy wetlands are supplied either by groundwater or by the clean runoff from woods and grasslands. I tell students that mosquitoes may check in, but they won't check out of a healthy wetland.

Wetlands that contain mosquito larvae are

considered unhealthy because they generally collect polluted waters from parking lots, roads, construction sites, cropland, pastureland, and lawns treated with pesticides and fertilizer. In many areas, mosquitoes are raised in small pools found in birdbaths, livestock watering tanks, old tires, and plastic toys left outside; on tarps; in sections of gutters that don't drain; and in buckets left around the home. I also find mosquitoes breeding in the remnants of drained and filled wetlands, that is, the bottom of drainage ditches, pools standing in partially drained fields, and ponds that livestock are allowed to access.

The origin of waters that enter a wetland is an important factor to consider when selecting a site that will not produce mosquitoes. Waters from undeveloped land are generally good, whereas runoff from urban land is generally bad. Avoid locating small wetlands in ditches that carry water from a football field or near parking lots used by students with old, leaky vehicles. In some situations, it may be best to build a wetland so that it receives no runoff; we call this a sky pond. You can expect a sky pond to maintain plenty of water in locations east of the Mississippi River, while west of the river it may be necessary to occasionally fill these systems from a garden hose.

Do Wetlands at Schools Have any Real Value?

In 2002, Beverly McDavid built a small wetland at Elliott County Middle School in the mountains of eastern Kentucky. I asked if the wetland is helping to achieve any science education goals at her school. Her reply:

The wetland transformed the schoolyard into a discovery zone, where students now explore the natural world and create a connection to it. The students love going to the wetland, and they learn so much more than they ever could from just using a science textbook. Inquiry investigations are easy for me to implement when using the wetland, and I try to incorporate the outdoors into my curriculum whenever possible. Studies show

that adding nature to your curriculum improves student learning, and I have seen these improvements. Our students have shown definite improvements in science knowledge as shown by our school's seventh-grade science scores.[4]

Once I asked, there was no stopping this response from McDavid, winner of a state Environmental Educator of the Year Award and recipient of a Disney Teacher Award:

Wetlands are a comprehensive educational resource—a living textbook, where students can learn about the water cycle, botany, food webs, soil, watersheds, and wildlife all in one place and firsthand. A wetland can also be used as a medium to teach non-science subjects such as art, math, English, and social studies. With so many things to study, the same wetland can be used as a teaching tool from kindergarten all the way through high school and beyond.

Margaret Golden worked tirelessly over a 3-year period and became the first educator in Lexington, Kentucky, to gain approval to build a wetland. I asked if this wetland has contributed in any substantial way to education at her school. Her reply:

Our wetland contributes in a number of ways to education at Beaumont Middle School. It has increased the number of classroom lab activities for students with real world relevance. It has increased opportunities for students to participate in independent study projects, such as science fair projects, which can be a real struggle for seventh graders. And, it's increasing opportunities for service learning projects, which builds self-esteem for students who volunteer hours with organizations like Beta Club, the National Junior Honor Society, and Scouting.[5]

Critics of building wetlands at schools claim they will eventually be ignored because of rigid curriculum requirements. However, the teachers who build them disagree. Beverly McDavid states:

The curriculum I use is aligned with state and national education standards. I am careful to follow approved curriculum and accountability guidelines. We are tested in science at the seventh-grade level, which puts a lot of pressure on me, not only to cover the areas on the seventh-grade program of studies for science but also to reinforce all areas in the core content for grade levels five through seven. I can use the wetland in such a large variety of science topics that it is one of the best resources I have. I will never stop using it![6]

Science teacher Ronetta Brown, who has built three wetlands at schools in Kentucky, says, "We'll never stop using these wetlands because they help us attain curriculum goals in an outdoor lab setting. Students are much more motivated to study in an outdoor setting than a classroom setting. Outdoor classrooms have become a critical piece of the entire school environment."[7] Margaret Golden offers:

There are few if any other ecosystems that can be placed in a tiny space and yield so much scientific information. Maybe grasslands, but they're difficult to establish with so many invasive species, and it's even harder to keep them from getting mowed. Many schools are land locked, and the wetland ecosystem provides a great deal of data for a relatively small investment.[8]

Margaret, Beverly, and Ronetta went to the trouble of preparing lists containing all of the state-mandated science education goals that the investigation of wetlands can help meet in grades 6, 7, and 8. Wetlands could help with at least four goals for each grade. One thing I know for certain: educators are quick to share what works for them. This probably explains why my hipboots are knee deep in requests from teachers for help with outdoor classrooms.

Hand-Digging Wetlands

People have been making wetlands by hand for thousands of years, often laboring for days to dig ponds we call wetlands to store drinking water, water livestock, operate steam engines, and even make moonshine. It may be necessary to create a

wetland by hand when it is not possible to access the site with heavy equipment, or when high-value trees are present and there is a risk of damaging them with heavy machinery. Of course, there are other advantages to using this technique that include low cost and plenty of aerobic exercise.

The two main factors important to creating a wetland by hand are suitable soils and compaction. Soil textures need to be high in clay or silt so they can be compacted to hold water when moisture is added.

To construct a wetland by hand, first a depression must be dug with a shovel. The soils in the depression must then be greatly compacted to hold water. Tree roots and topsoil must be removed from the depression before it is compacted, as they can create channels for water to leave the wetland. It is possible to obtain adequate compaction by using a heavy steel spud bar, such as that used for fence construction, to repeatedly tamp the bottom of the wetland. This is a lot of hard work, even for a small area. The chance of success is increased by placing soil in the depression in layers 2 to 3 inches thick and thoroughly compacting between each layer.

One can also obtain suitable compaction by repeatedly driving over a site with a rubber-tired vehicle or by using a hand-operated motorized compactor, which is available at most equipment rental businesses.

Using a synthetic liner can greatly improve the chance of success when establishing a wetland by hand. Fortunately, the cost of a liner for a small wetland is generally within the budget of many homeowners.

It can be tempting to use a pre-formed plastic tub to create a small wetland by hand, but be sure to avoid the types made with steep sides. Water levels will drop in any wetland, and when they do, amphibians, reptiles, and small mammals can be trapped inside these steep-sided structures. Using soils and a flexible synthetic liner allows you to place gently sloped sides around the wetland that will provide a reliable escape path for wetland animals.

(left) Small ephemeral wetland resulting from a hand-dug pit in the 1830s, in the Daniel Boone National Forest in Bath County, Kentucky. Only a small number of the hundreds of pits originally dug for the removal of iron-rich clay hold water today.

(above) Ephemeral wetland dug by hand with a mule and scoop in 1942 to provide water for a steam-operated sawmill in Rowan County, Kentucky.

Wetland Construction and Synthetic Liners

A synthetic liner allows you to build a small wetland from soils that are too porous to hold surface water. Sandy and rocky soils found where the water table is far below the surface are generally passed over when searching for wetland establishment sites. Using a liner can be a reliable and cost-effective means to establish a wetland on permeable soils when there are no other feasible locations available for construction.

Using a liner may be the only way to establish a wetland near schools or nature centers. Soils near these buildings often consist of construction fill that includes chunks of concrete, rock, asphalt, and topsoil. Construction fill is porous and cannot be reliably shaped into a depression that will hold water.

It is now possible to build a wetland at most any school by using a synthetic liner. To find a good location for establishing a wetland, I recommend walking around the school grounds and looking for a level area approximately 60 feet in diameter, within a short, 5-minute walk of the school. Wetlands can be built farther from the classrooms, but they are less likely to be used by teachers because of the time it takes to reach the site. Don't let the presence of a few small trees or shrubs keep you from selecting an area, as these can easily be transplanted by the heavy equipment on the day of construction. Avoid obvious conflicts by steering away from ditches lined with cement, sites containing storm sewers, and areas used for sports practice or play.

Denote where a wetland can be built by using wire flags and a tape measure to mark either a circle 40 feet in diameter or an oval 30 feet wide by 40 feet long. It is important to keep the marked edge of the wetland 10 feet or more away from buildings, trees, parking lots, fences, or utility poles as space will be needed for equipment to operate and for students to walk around the wetland.

Once you have marked a suitable location, contact maintenance personnel to check for the presence of buried utilities such as electric, gas, phone, and fiber-optic lines, as well as water drains and storm sewers. Be ready to change locations if utilities are buried on the site. For additional safety, ask that the location of all buried utilities be marked within 50 feet of the proposed project, and later alert the equipment operators of their presence before excavation begins.

Liners made of PVC (30 mil or thicker) and EPDM (synthetic rubber, 45 mil or thicker) work well for wetland construction. The need to use aquatic-safe/fish-grade liners cannot be overemphasized. Most synthetic liners used for roofing, landfills, and tarps are treated with fungicides and algicides. These chemicals are toxic to aquatic life. I have examined a number of wetlands built at schools in eastern Kentucky where educators used locally purchased synthetic liners with dismal results. Granted, these wetlands held water, but the waters were devoid of life. Netting and seining these sites failed to reveal the presence of any living organism. Mammals and birds that drink the water from these ponds almost certainly suffer from the chemicals these liners are treated with. Jim Enoch, engineer with Just Liners Incorporated, claims that it is almost impossible to purchase an aquatic-safe liner locally in most communities as the standard for the industry is to sell treated liners.[9]

Liners can be ordered from the factory in about any size. I've had the best success using liners that measure 40 × 40 feet (to construct a circular wetland) or 32 × 40 feet (to construct an oval wetland). These sizes are best suited for building wetlands from 12 to 20 inches deep with gradual slopes. Because manufactures commonly seam synthetic liners from materials on 8-foot-wide rolls, you may save money and reduce waste by ordering a liner in increments of 8 feet.

Because of their weight, a 40 × 40 foot liner is about the largest a crew can be expected to install without using specialized equipment for transporting and positioning. When building a wetland, I'll enlist the help of six or more strong adults or a class of thirty students for moving, unrolling, and positioning the liner.

A landfill may be a good business to contact for the possible donation of a synthetic liner for building a wetland at a school. These operations use great quantities of synthetic liner material to seal acres of land and may be willing to donate a piece of the material and possibly help with construction.

Using a thick, durable, synthetic blanket, known as a geo-textile pad, to protect both the top and bottom of the synthetic liner is always a good investment. Years in the future, someone unbeknownst to the builder might use a soil probe or a shovel to sample a constructed wetland, accidentally puncturing an unprotected liner. Geo-textile fabric will help guard the liner from puncture and from possible damage caused by burrowing crayfish, gophers, tree roots, and deer hooves. Geo-textile fabric is generally available for purchase from the same businesses that sell synthetic liners and can be ordered in pieces the same size as the liner.

Before construction begins, you should measure how much the ground slopes on the location where you want to build a wetland with a liner. Should the elevation from the upper to the lower edge of the work area exceed 6 inches, or 1 percent, I recommend hauling in additional soil to use for covering the liner. Here's what happens when you build a wetland with a liner on a slope steeper than 1 percent. All the soil removed to create the depression for the wetland is used to form a dam with gradual slopes. This means that no soil is left over for placing a layer over the liner. When constructing a wetland away from buildings out in an open field, you can generally "borrow" soil from an area near the wetland to cover the liner. However, this rarely is possible on school grounds, where the removal of soil from other areas can cause erosion or damage parking lots, planted trees, and underground utilities.

It's surprising how much soil is needed to cover the liner. For example, when building an oval-shaped wetland measuring 32 feet × 40 feet on a slope that changes 1 percent or less, approximately 30 cubic yards of soil is needed to place a 6-inch layer over the liner. Even more soil is needed on steeper ground. Should the elevation change by 6 to 12 inches, an additional 10 cubic yards of soil is needed. Don't be concerned about ordering a little too much soil for your project, as excess soil can always be used to blend the dam into the surrounding landscape.

Construction and landscaping contractors are good sources for soil that is not sandy or rocky. You should not use soil with sharp rocks, as they can puncture the synthetic liner. Soil containing round rocks can be used to cover the liner, but rocks with sharp edges should be removed, generally by hand during construction. Topsoil is not recommended for covering the liner. It can be expensive and may contain roots and branches that can puncture the liner. Topsoil is also difficult to spread and may contain unwanted nonnative, invasive plants. The soils you order should be dumped adjacent to the wetland construction site, and they may be delivered weeks before the project begins.

A dump truck should be used to haul the additional soil needed for a project. Dump trucks are built in different sizes and are classified by the number of axles supporting the dump bed. A single-axle dump truck will haul approximately 8 cubic yards of soil, a double-axle (tandem) 13 cubic yards, and a tri-axle 18 cubic yards.

Before you build, a test hole should be dug to see if there is enough soil on the site to form the dam for the wetland. Occasionally, a location may be found on top of rock, so soil cannot be removed to create a depression for the wetland or to shape a dam. If you can dig a small-diameter, 24-inch-deep hole in the center of the proposed site, there should be enough soil to create the depression and build the dam. You'll need to order approximately 50 cubic yards of soil to build a 32 × 40 foot wetland on a site that is on top of bedrock. One-half of the soil is needed to form a dam, while the other half would be used to cover the liner.

Heavy equipment is needed to create the depression for the new wetland. After building more than sixty wetlands with liners, I've found that a small excavator, one that also has a blade along with rubber-covered tracks, is the ideal machine to use when building at a school. These small excavators are commonly owned by contractors and can be hired at a reasonable price. Because they have rubber-covered tracks, they can be operated on grass, blacktop, and concrete surfaces without causing damage, and they are small enough to pass through gates. They also work well in small spaces adjacent to buildings and will not disturb large areas around the work site as occurs when using a dozer.

A small front-end loader on tracks, commonly called a track-steer, also works well for building a wetland with a liner. The main drawback is that it is not as effective as an excavator for digging and shaping the wetland. This means that more shoveling and raking will have to be done by hand. I've also used a dozer to construct wetlands with liners at schools with success. Using a backhoe is problematic. Backhoes are large and require much room to operate, and their floppy attachments can hit walls and trees and get hung up in overhead wires and branches.

Before construction, mark the shape of the wetland you would like to build on the ground with wire flags and orange spray paint. When using a 40 × 40 foot liner, mark a circle whose diameter is 40 feet; when using a 40 × 32 foot liner, mark an oval approximately 40 feet long by 32 feet wide. Set up a level on a tripod near the construction site so it can be used to measure depth and slope during construction.

Begin construction by asking the operator to dig a hole that is 2 feet deep in the center, with gradual, tapered slopes that run out at the marked perimeter. Let the operator know that the depression should be shaped like a large satellite dish, that it does not have a flat bottom, and that soils can be left rough around the edges until the liner is in place. It is also good to say that you would like the liner covered with approximately 6 inches of soil and that it should be no deeper than 18 inches when finished. The top edge of the depression should be level, just like the rim of an extra-large cereal bowl set on the ground. Use the level and rod to measure the depth of the depression in relation to the lower edge of work site when digging begins. The depth of the hole in the center should be checked a number of times during construction, as operators who are new at building wetlands with liners may have difficulty gauging the correct depth without assistance.

Small ephemeral wetland constructed in 2003 with a synthetic liner at the Minnesota River Valley National Wildlife Refuge Visitor Center in Minneapolis.

When the operator believes that the excavated depression is ready for the liner, check the depth of the depression and the elevation of the surrounding rim before setting the liner in place. Start by marking the center of the depression, recording its depth, then going uphill 2 feet and marking this elevation around the inside edge of the depression with wire flags. Now use a long, flexible tape measure to determine the distance across the depression from flag to flag. At no place should the distance between opposing flags be greater than the length or width of the liner. Rarely is a depression constructed the correct size and depth the first try, so expect to dig deeper, make the depression longer, or fill in part of the hole to obtain the right size for the liner. It is generally faster to make the depression a little too small at first and then expand it, rather than make it too large at the outset. Placing the liner in the depression is like burying a children's swimming pool in the ground: the top rim must be level so water does not leak over an edge.

In rocky soils, it is a good idea to travel over the completed basin a number of times with the heavy machinery before placing the pads and liner. Repeated passes will break up large rocks and help push sharp, protruding edges into the ground, reducing the possibility of puncture.

The liner and geo-textile pads should be anchored along the top edge before covering with soil. The anchors will keep the liner from being pushed down into the depression into a hopeless mess. I find that 12-inch-long landscape spikes with washers work well for anchors. These spikes puncture the liner and pads with ease and can be driven into rocky soils with a small sledgehammer. I place the anchors about 3 inches below the top of the liner, at intervals from 12 to 18 inches apart.

Once the liner and geo-textile layers have been placed in the depression, the top edges should be trimmed to match the desired depth of the wetland. Use spray paint to mark spots indicating where the top of the liner should be trimmed.

This requires the cooperation of two people: one to look through the level, and the other to walk around the wetland with the rod and spray paint. The person operating the level has the sole job of informing the person holding the rod when the elevation being read is 2 feet higher than the bottom of the depression. The person with the rod begins by standing near the inside top edge of the liner. The rod is moved slightly up or down the slope until the number being read is 2 feet higher than the bottom of the wetland. A small dot is painted on the top of the synthetic liner at the base of the rod to mark the elevation. The rod holder then moves another 3 feet along the top edge of the liner and repeats the process. This is continued until a series of dots, all painted at the same elevation, mark the desired top edge of the liner.

The next step involves anchoring the top edge of the liner and pads with spikes and washers every 18 inches, using the circle of painted dots as a guide. Once the top edge is anchored, the excess liner and geo-textile fabric remaining above the spikes should be trimmed and removed. Sharp utility knives should be used to cut all three layers at one time, leaving 3 inches of material or less above each anchor. Leaving more material above the spikes makes it difficult to hide the top edge of the liner, creates a tripping hazard, and unnecessarily uses soil needed for covering the liner. The resulting raised rim prevents runoff from entering the wetland, is unsightly, and gives the appearance that someone purposely built a large donut around the wetland. A layer of soil should be placed over the liner once it is trimmed.

There are many benefits to covering a liner with a layer of soil that is at least 6 inches thick. The soil blanket protects the liner from being punctured by horse, cattle, and deer hooves and shields the liner from ultraviolet damage caused by the sun. Soils also provide a medium for plant growth and a substrate into which amphibians and reptiles can burrow to escape predators, and for hibernation. Use hand rakes to place a gradual slope around the inside edge of the wetland, which creates various water depths that are im-portant to maintaining plant diversity and to animals entering and leaving the wetland. Liners not covered with soil become very slippery around the edge, making it easy for students to slip and fall during investigations.

It is important to keep heavy equipment off the liner when covering it with soil. I've experienced wetland construction failures when a dozer has traveled over parts of the liner. Several of the wetlands I have built with a liner failed to hold water, most likely because the liner was moved or torn beneath the ground by the dozer or backhoe, a problem not visible during construction. I will use heavy equipment to push soil up to and over the edge of the liner, and then only use hand rakes and shovels to spread soil over the liner. I have had 100 percent success in creating wetlands with a liner when the liner has only been covered by hand labor. Make sure that adequate soil covers the top edge of the trimmed liner so that students will not trip over an exposed edge at a later date.

Use care to make gradual slopes around the inside edge of the constructed wetland so students are less likely to slip and fall. A gradual slope also holds up better to foot traffic and is less likely to break away and erode. It is also important to place a gradual slope around the outside edge of the wetland so that it can be maintained by a riding lawnmower if needed. A gentle slope from 10:1 to 20:1 (10 to 5 percent) works well around wetlands intended for educational use.

During construction, if it appears that there will not be enough soil to both cover the liner and create gradual slopes around the wetland, lower the profile of the wetland by excavating a depression that is longer and wider than the size of the liner, and another foot deeper than the depth you had planned. Installing the liner in the bottom of the deeper depression will make more soil available to place over the top of the liner. This change results in a more sunken wetland that has longer slopes leading down to the edge of the water. The elevation of water in the wetland will still be controlled by the height of the top edge of the liner in the depression, not by the higher rim of soil surrounding the wetland.

When covering the liner with gravel or sandy soils it is best to place a gradual slope, from 10:1 to 20:1, around the inside edge of the wetland. Otherwise, sand may slip off the liner on steeper slopes, exposing the material to damage from the sun and puncture by animals.

We regularly plant aquatics in the 6-inch layer of soil covering the geo-textile pads and liner in the new wetland. It is quite common to encounter the geo-textile pad when digging holes for the plants, which is no reason for concern as the fabric is strong enough to resist cutting with a shovel.

Here's a list of supplies and estimated prices for constructing a circular wetland at a school with a synthetic liner:

1. Synthetic liner: PVC, 30 mil or thicker, aquatic-safe grade, 40 feet × 40 feet (1,600 square feet) × ($0.40/square foot), quantity 1 = $640.00.
2. Geo-textile pads: 8-ounce weight or heavier, for protecting the top and bottom of the liner, aquatic-safe grade, 40 feet × 40 feet (1,600 square feet) × ($0.30/square foot) = $480.00/each, quantity 2 = $960.00.

   The PVC liner and geo-textile pads may be ordered factory direct from

   Fabseal Industrial Liners, Inc.
   42404 Moccasin Trail
   Shawnee, OK 74804
   1-800-874-0166
   http://www.fabseal.com

   Just Liners, Inc.
   35507B Clearpond Road
   Shawnee, OK 74801
   1-888-838-4017
   http://www.justliners.com

3. Nails and washers: You'll need pairs to anchor along the top edge of the liner and pads for every 18 inches. Purchase 12-inch-long galvanized landscape spikes, sold by the pound, approximately 3.5 spikes/pound, $1.30/pound = $0.37/spike (purchase 85 spikes @ $0.37/spike = $31.45). Washers, one per spike, 7/16-inch center hole, $0.12/each or $3.69/box for a box of 100 (purchase 1 box @ $3.69/box = $3.69).

4. Native grass seed mix: Sow a mixture of native grasses on exposed soils that are above the water level for erosion control and to reduce the potential for invasive plant colonization. Wild rye, big bluestem, little bluestem, switchgrass, and Indian grass work well for this purpose. Order 2 pounds of native grass seed mix to sow; you may need to mail order this seed and can find a native plant nursery serving your area on the web. Expect to pay up to $25.00/pound, or a total of $50.00 for the seed.

5. Straw: Purchase nine bales to use for mulch. All areas of exposed soil, except for the bottom of the wetland, should be covered with a layer of straw to reduce erosion and to increase plant survival. Do not use hay, as it contains more weeds that may have to be controlled at a greater expense at a later date. Plan to mulch up to 8,000 square feet; one bale of straw will cover an average of 900 square feet. Straw can be purchased at a local farm supply store for $3.75/bale; nine bales needed = $33.75.

6. Wheat: Purchase one 50-pound bag of winter wheat for the project. Wheat provides for the rapid control of erosion and should be seeded immediately after construction. Wheat can be purchased at a local farm supply store for about $8.00/bag.

7. Aquatic plant seed: You may wish to order a small package of native wet-meadow seeds to sow around the water in the new wetland. A 0.25-pound bag of mixed seed should be ample for the job. You'll probably need to mail order this seed and should be able to find a native plant nursery that serves your area on the web. One package = $125.00.

8. Soil: Order the necessary quantity of rock-free soil by the dump truck load when building on a slope or on rocky ground. Expect to pay up to $10.00/cubic yard for the soil, with most projects requiring three loads, or 30 cubic yards. Total = $300.00.

9. Heavy equipment: Hire a contractor with an excavator, dozer, or track-steer to move soils for constructing the wetland. Plan on paying for up to 7 hours of heavy-equipment work to complete the wetland; however, some operators may charge a minimum of 8 hours. Estimated cost ($80.00/hour) x (7 hours) = $560.00.

10. Brightly colored wire plastic flags: 12 to 18 inches long, quantity needed = (20) × ($0.10/each) = $2.00.

11. One can of fluorescent orange spray paint: Purchase the can type that is used by contractors to mark the ground and can be sprayed upside down = $6.00.

12. Level, tripod, and rod: Heavy-equipment operators are generally willing to provide this equipment at no additional charge.

13. Rakes (six) and shovels (three): Perhaps your friends can bring these with them when they volunteer to help.

14. Sledgehammer: It's best to have two small sledgehammers available for driving the landscape spikes into the ground. Again, perhaps these can be borrowed.

15. Knives: Sharp razor-blade knives, the plastic type with break-off disposable blades, work best for cutting the top edge of the liner and the geo-textile pads. Purchase three @ $3.00/each = $9.00.

Visitor Center Wetlands

Forest Service Interpretative Specialist Evelyn Morgan worked with a number of volunteers to build a short trail around the new Forest Service Visitor Center overlooking Cave Run Lake. She later obtained a grant to build an observation blind along the trail beneath a canopy of large white oaks. I suggested to her that we build a couple of ephemeral wetlands near the blind to attract wildlife, mentioning that these would also be great places to sample with school groups. Evelyn liked the idea but made me pinkie-swear that no trees would be removed for the project.[10] The wetlands would have to be built in small gaps be-tween the oak trees, where there was no room for heavy equipment.

I tested the soil on the construction site with a handheld soil auger and found a mixture of silt loam and gravel. I knew that synthetic liners would be needed to hold water as it would be difficult to remove the large tree roots and to pack the mixed soils. Fortunately, I heard that Boy Scout Todd Watts was interested in directing a project to help him attain the rank of Eagle. We decided to work together to build the wetlands so that both of our objectives could be met.[11]

Fifteen Boy Scouts and five adult leaders showed up early one Saturday to construct the wetlands. They began by raking leaves and branches off the two sites. Todd's father then used a rototiller to loosen the ground. The boys grabbed shovels to dig large holes 18 inches deep, taking about one-half of the soil they removed and placing it along the lower edge of each depression to make the sites level. They saved the other half of the soil in piles for spreading over the liner later. Rakes were used to smooth the depressions and to make gradual slopes along their edges. Fire rakes with sharp blades were used to cut exposed roots in the holes. Each cut root end was covered with soil to keep it from puncturing the liner. Using scissors, they cut the synthetic liners slightly larger than the size of each depression. The boys placed the liners in the holes, covering them with 4 to 6 inches of soil. Several of the Scouts took great pains to remove rocks and sticks from the soil so they would not puncture the liner as it was covered. The Scouts then dragged logs and branches into the ponds, seeded exposed soil to winter wheat, and spread a layer of straw to control erosion.

The small wetlands filled with water by early winter and now generally go dry by late fall. They are only 8 inches deep and measure 6 feet wide by 8 feet long. Over the years, students have used nets to capture wood frogs, green frogs, gray tree frogs, spring peepers, American toads, spotted salamanders, dragonflies, damselflies, water boatmen, and backswimmers in the wetlands. Todd Watts earned the rank of Eagle and is now attending college.

Flourishing Flora

Wetlands are generally defined as areas having hydric soils, the presence of water (either standing on the surface or saturating soil during all or part of the growing season), and plants adapted to living under saturated conditions. To help assess if a constructed wetland is a success, A. E. Plocher and J. W. Matthews ask the key question, "Does the site have dominant hydrophytic vegetation?" They also suggest assessing functional problems regarding whether or not the vegetation is unacceptable by listing these two assessment criteria: (1) exotic or weedy species should not be among the most dominant species in any vegetation layer, and (2) not more than 50 percent of species present should be exotic.[1]

The growing and marketing of plants for use in wetland mitigation projects appears to have become a successful business in the United States, as the U.S. Army Corps of Engineers identified sixty-two nurseries, primarily in the Northeast, specializing in the sale of plants for wetland projects.[2] Examining the catalogs from several of these aquatic plant nurseries shows that plants for a wetland project could easily cost thousands of dollars.

The successful establishment of desirable plants in wetland restoration and creation projects has presented great challenges and expense to many involved in the profession. My observations show that establishing aquatic plants in a wetland project largely depends on first creating a desired hydrologic regime. If the site does not hold water as planned, it will be difficult, if not impossible, to establish obligate or facultative wetland plants.

One does not always have to spend money on aquatic plants for a successful wetland restoration project. April Haight examined plant colonization of constructed wetlands in the Licking River watershed within the Daniel Boone National Forest and found that fifty obligate wetland and facultative wetland plant species grew in constructed wetlands within 5 years of establishment. She found species richness to be greater in constructed wetlands than in the natural reference wetland examined. The lower species richness in the natural wetland was believed to be caused by more stable water levels and a dense growth of *Nuphar advena,* which was not present in the constructed wetlands. Facultative wetland and obligate wetland plants ranged from 67 to 87 percent in the constructed wetlands studied. The lowest concentration of wetland plants occurred in constructed wetlands that were drawn down during the study period.[3] These numbers are impressive, considering that no planting of wetland species occurred after construction and that topsoil was saved and spread on the bottoms of only three of the five constructed wetlands examined.

I have experienced greater success in establishing wetland plants, and at a lower cost, by planting containerized species such as sedge, bulrush, buttonbush, hibiscus, and cardinal flower after wetland projects have filled with water. Waiting until a location obtains its planned hydrology makes it easier for planting crews to avoid placing species where they would later be inundated and killed. One factor to consider is that it is often too dry to plant in the summer and fall just after a wetland is completed, and when planting is done during drought conditions one can expect high mortality. A number of authors describe using irrigation systems to help increase the survival of plantings made in a dry time of year. Although irrigation will work, its success depends on the commitment of personnel and on funding that is adequate to maintain the system long enough for precipitation to return and ensure survival.

A common trap to fall into with wetland restoration is to attempt to produce a great diversity of aquatic plants within a year or two after construction. Expecting to find uncommon and showy species such as hibiscus and iris growing in a wetland soon after it's built can be unrealistic, as well as expensive. Jim Lempke, arborist at the University of Kentucky Arboretum, teaches that it is much more important to focus on establishing grasses and sedges in a new wetland and then later work on returning the less common and showy forbs.[4] Visit a natural wetland and even take vegetation plots in one, and you'll find grasses, sedges, and rushes to be the most common wetland plants. It makes good sense to first establish the more common native species so that nonnative invasive species are less likely to colonize exposed soils.

Lempke offers this good advice at the workshops he teaches about restoring native plants and controlling invasive species: "You can't solve all the problems of the world," he'll say to make it clear that one should not be overwhelmed by the task of returning native plants to a restored wetland or prairie. He also urges patience when he advises, "Don't expect instant success; it may take years for wildflowers to bloom." In following his own advice, Jim often uses seed to establish native plants, as collecting seed from native populations has much less of an impact than digging and moving green material.

After attending one of Lempke's workshops, I began collecting plants from donor populations to use in wetland projects. Finding plants that were available at no charge was really quite easy with all the wet roadside ditches, edges of farm ponds, and construction sites in my community.

The presence of nonnative fish, such as carp, can also influence plant diversity in a wetland. I

The cardinal flower, one of the showiest plants that grow in moist wetland soils.

manage a number of restored wetlands near Cave Run Lake, a flood-control reservoir in the Daniel Boone National Forest. Every couple of years the lake will flood, and can rise 26 feet or more above summer pool for months, inundating restored wetlands. As the floodwaters recede, carp are one of the fish species left behind in wetlands. The carp soon eliminate all submerged aquatic plants from the wetlands such as pondweed, eel-

Three emergent wetlands built with permanent waters near Clearfork Creek in the Daniel Boone National Forest are graced by the fragrant and beautiful white water lily.

grass, and *Najas minor*. These carp introductions are the main reason we place drainpipes in the affected wetlands, so that they can be easily expunged after a flood pool event.

One of the problems I have faced in wetland restoration is the colonization of a dense growth of cattails, which can become a weedy species in some locations. Within 5 years of establishment, cattails can develop so thickly in new wetlands that it is difficult to find open water. Cattail seeds apparently blow into recently completed wetlands from adjacent wetlands and may also be present in the topsoil spread over the bottom of recently completed wetland projects. Cattail seeds germinate in exposed moist soils.[5] Most of the wetland construction work I complete is done in late summer and early fall, which also corresponds to the time of year when cattail seeds are dispersing. When moistened by rain, the soil in the bottom of these new wetlands provides ideal conditions

for cattail germination. The cattails will continue to grow best in those wetlands that receive full sunlight, are less than 3 feet deep, and do not dry completely each year.

Fortunately, I have observed that muskrats have a large impact on cattails and can even eliminate them from wetlands. Since 1989, I have worked to build a complex of ninety wetlands in the Beaver Creek watershed in Menifee County in the Daniel Boone National Forest. Wetlands restored and created near Beaver Creek include ephemeral, emergent, wet-meadow, and forested types, ranging from 0.1 to 6 acres in size. Each year I inspect these wetlands to identify maintenance needs and typically find that from two to five wetlands are choked with narrow- and wide-leaf cattails. Similar inspections completed the following year often find cattails all but eliminated from the wetlands dominated the year before. Other changes noted are that these more open wetlands will now have abundant muskrat signs such as feeding activity, runs over the dam, and houses built of cattails.

Paul Errington described the relationship between increases in muskrat density and corresponding decreases in cattail abundance that follow in wetlands.[6] Unfortunately, muskrats do not appear to be present in all areas where we have restored wetlands in the Daniel Boone National Forest. For instance, in the Clearfork drainage of Rowan County where muskrats are absent, cattails continue to colonize wetlands until actions are taken to decrease their density. To date, we have removed water from wetlands during the growing season, mowed cattails when wetlands are dry, and even scraped cattails out of dry wetlands with a dozer, all with limited success. However, cattails were all but eliminated from one wetland after beaver blocked its water-control structure, raising water levels for 1 year.

Private landowners are known to purposely construct ponds that are deep with steep sides to prevent cattail invasion, desiring more open water for esthetics and easier fishing over vegetation diversity and wildlife habitat. Their actions effectively discriminate against cattails but also produce bodies of water unlike natural wetlands that once occurred in many areas. In the future, in an effort to outcompete germinating cattails, I intend to seed high densities of winter wheat and millet, both noninvasive annuals, in the bottom of recently completed wetlands. But perhaps it would be better to return muskrats to these problem areas, so that they can make better use of the resource.

Should you decide not to wait and see if muskrats will eventually control cattails in a wetland, herbicide application provides an effective means of reducing a dense growth of the plant. Wildlife Biologist Terry Moyer, who manages Richardson Wildlife Foundation lands in Illinois, regularly uses an aquatic-approved formulation of the herbicide glyphosate to control cattails in wetlands.[7] I have also found that cattails can be controlled in constructed wetlands by applying the herbicide and surfactant with a handheld sprayer. Unfortunately, the chemical is nonselective and will kill all other plants that are sprayed.

Completely drying a wetland is a good way to shock and kill unwanted aquatic plants, such as cattails, without using herbicide. Terry Moyer has found an effective way to dry both the surface and subsurface of restored wetlands in order to fight unwanted aquatic vegetation. Since the majority of wetlands that Terry builds are located on top of buried drain lines, he finds where the drainage structures pass beneath the dam being constructed, then uses a dozer to block the smaller-diameter lateral lines. He searches for the largest-diameter drain line, which generally represents the main line. Instead of crushing the line as one normally does during wetland construction, he intercepts it with a water-control structure that is placed upright in the dam. Later adjustment of the plates in the water-control structure will thus change the elevation of both surface water and groundwater in the finished wetland. He replaces a section of the porous drain line with solid-wall plastic pipe to reduce the possibility of water seeping under the dam. While examining wetlands at the wildlife foundation, Terry showed me where he set one of these water-control structures 12 feet in the ground in order to tie into an 8-inch-diameter clay tile line.

Establishing wetlands often involves building levees and disturbing soils that will not be inundated. The exposed surfaces left after construction can be sloped and subject to erosion. Obligate wetland plants and many facultative wetland plants have little chance of survival when planted on these well-drained soils in wetland project areas. The hot, dry surfaces present during the construction season are also unlikely to naturally vegetate in late summer and fall. My main concern with these exposed soil surfaces has been to control erosion and to protect the wetland investment. Even the most natural-appearing restored wetland basin can be severely damaged by flood conditions without the protection of vegetation. I have seen individuals, as well as various agencies, spend thousands of dollars in building a wetland only to skimp on essential vegetation needs above the water surfaces.

Just as soon as heavy equipment finishes grading an above-water area on a wetland construction site, it should be seeded to winter wheat at a rate of 80 pounds per acre. I have found that this immediate sowing often allows us to capture the unexpected benefit of infrequent summer and fall rains, causing the wheat to germinate within 3 days. Wheat provides rapid protection against washing and is a noninvasive annual species that has worked well for projects completed in Alabama, Kentucky, Maryland, New York, North Carolina, and South Carolina. It is best to sow wheat before the first rain falls on a project, as germination rates drop after fine-textured, unpacked surface soils have washed away. I have had greater success using wheat than annual rye or oats. Another advantage is that wheat is generally less expensive to purchase than other species.

Almost as important as seeding wheat is the application of a layer of straw to the same exposed above-water soils on a wetland project. A layer of mulch greatly increases the germination of seed and survival of seedlings on a wetland levee. Those individuals who work alongside me on wetland projects place a high priority on spreading mulch on exposed surfaces before the first rain falls on a wetland project. The straw alone provides protection from the washing that can occur in a downpour when raindrops impact bare soil. I have seen wheat germinate and grow beneath a blanket of straw even under drought conditions. Morning dews and residual soil moisture can provide the water needed for germination in late summer and fall when rainfall is at its lowest.

Spreading a layer of topsoil on a completed dam can also help vegetation take root. In areas with acid subsoil, adding a layer of topsoil may be the only way to get vegetation to grow on slopes of newly constructed dams. Dennis Eger of the Indiana Department of Natural Resources always covers new dams with up to 6 inches of topsoil and finds this practice to be most helpful when working to stabilize exposed soils in a floodplain.[8]

A number of individuals also sow native prairie grasses on wetland dams and borrow surfaces at the same time wheat is applied. Switchgrass, big bluestem, and little bluestem appear to grow well on wetland dams. These species can take over protecting a dam from washing after wheat has grown and died.

CHAPTER SEVENTEEN

Fixing Failed Wetlands

Walking up to a constructed wetland that was designed to hold water and finding it dry would give anyone a sick feeling, especially the wetland builder. Knowing whether or not a wetland is functioning as designed can be a challenge and may take as much time to figure out as building a new wetland. A. E. Plocher and J. W. Matthews describe a reasonable means for determining wetland success known as a "Wetland Assessment Procedure" that contains eight general attributes for evaluating the functional success of a constructed wetland: wetland status, functional problems, realism, floristic quality, size, and landscape setting; wetland type and water quality are suggested for use when the wetland is expected to meet specific requirements.[1]

They state in their book:

The first and most important consideration is whether the site is a wetland. The generally accepted procedure is to use the U.S. Army Corps of Engineers' criteria for identifying jurisdictional wetlands (Environmental Laboratory 1987). A wetland must support hydrophytic vegetation, hydric soils, and wetland hydrology. There is no quality rating for this category. If the site is a wetland it passes, if not it fails.[2]

The most common reason I have seen that wetland projects fail is that they do not develop wetland hydrology; simply, they do not hold water long enough for hydric soils and plants to develop. In their textbook on wetland ecology, William Mitsch and J. G. Gosselink state that "hydrology is the most important variable in wetland design. If the proper hydrologic conditions

are developed, the chemical and biological conditions will respond accordingly."[3]

It should be easy to make a wetland where one used to be located; regrettably, this is often not the case. I have met and talked with a number of people who have built wetlands in the United States and Canada and have found that close to one-half of their wetland projects fail to maintain planned hydrology. The person who builds a failed wetland often faces embarrassment and a lack of support for continued restoration work.

Fortunately, most failed wetlands can be repaired, and generally for less than their original cost. However, most people walk away from a failed wetland, shrugging their shoulders while mumbling that nothing more can be done. Some go on to say that these failures prove that we don't know enough about wetlands to build them and should think twice before trying. The following tales describe how individuals have gone about repairing wetland projects that lacked hydric soils, aquatic plants, and desired hydrology.

Leaky Water-Control Structures

Many constructed emergent and forested wetlands have a water-control structure attached to a pipe that passes under the dam. Water-control structures are a common source of leaks and should be examined carefully when attempting to determine why a wetland does not hold water as planned. A site should be visited after a heavy rain or during a time of year when water should be present. Attempt to locate the water-control

(top) A failed constructed wetland due to damage to this flashboard riser caused by rust and gunshot holes near Cave Run Lake in the Daniel Boone National Forest in Kentucky.

(bottom) Emergent wetland repaired 15 years after construction by replacing its leaking flashboard riser water-control structure with gasketed PVC pipe at Land Between the Lakes in Kentucky.

structure or pipe that passes through the dam near the deepest part of the wetland. Check for leaks caused by poor-fitting joints, cracks, loose-fitting boards, and even bullet holes. Even a slight leak in a water-control structure on a wetland with a small watershed can drain a site in a couple of weeks. Suspect a leaky water-control structure if a failed wetland has a history of holding water.

The flashboard riser is a commonly used water-control structure for wetland projects and is often responsible for wetland failure. Consisting of an upright culvert cut in half, with slots for placing removable boards, the structure is prone to leaking and is generally only effective on those sites with a large watershed and a constant inflow of water. It is practically impossible to keep the flashboard riser from leaking, even though many have tried by using copious tubes of caulk in their attempt to seal around loose-fitting boards. These structures rust, usually along their seams; rusty seams will eventually leak, and these leaks can be difficult to detect. Replacing a steel flashboard riser with a tight-fitting PVC water-control structure with gaskets can often repair a failed wetland.

Flashboard risers are typically attached to corrugated steel culverts that then pass under the dam. These steel pipes may also have loose-fitting joints that can cause water to leak from the wetland, resulting in failure. I have seen where steel pipes were surrounded by rock and not packed with clay or silt loam. This rock left around a pipe will act like a drain to remove water from the wetland.

Fiberglass water-control structures that use plastic panels with synthetic gaskets for sealing gates, flaps, and boards became widely used by wetland managers beginning in the 1980s. These structures work well when first installed, but one brand, over time, has developed serious leaks. The main problem is that the gaskets will compress and fail to retain the resiliency needed to seal out water. Leaks in these structures can be difficult to detect and are sometimes only visible after a heavy rain. Loose-fitting flaps that are easy to remove indicate that the structure is leaking and should be replaced.

Holes in the Dam

A wetland with a long history of holding water can suddenly fail when a muskrat burrows a hole in the dam. The 4-inch-diameter hole a muskrat can make will drain a wetland quickly. Muskrats make dens by two means: one, by constructing above-water homes of aquatic vegetation with an underwater entrance; the other, by tunneling underwater into a bank and then excavating a living chamber above water, yet underground. Finding a muskrat hole that is allowing water to pass through a dam involves walking the dam in three passes: the first along the inside edge, the second on the top, and the third along the backside. Muskrat holes can be difficult to spot, requiring great concentration by someone with sufficient time to devote to the task. Some holes create channels through the dam, while others enter the dam part way and are barely visible under the water. Not every muskrat burrow is a problem; only those that allow water to pass through the dam have to be repaired. Muskrat holes can be tested by muddying up the water in the burrow on the inside of the dam, then watching to see if a stream of silt moves through the dam. Water should be removed from the wetland before muskrat damage is repaired, as it is difficult to pack the soils in a burrow when water is gushing in. A backhoe works well for digging out holes and packing them with soil, as most holes are too deep to fill with a shovel. Although a dozer can be used for these repairs, it exposes more soil, which in turn, needs to be vegetated, and therefore it increases costs.

Muskrat burrows can be hard to find. I worked to restore a 3-acre wetland near Beaver Creek in Menifee County, Kentucky, in 1991 by building a low dam of silt loam soils to trap runoff. The wetland held water as designed for 9 years, attracting a diversity of aquatic plants and animals. While sliding across the ice of the wetland to place straw in a Canada goose nesting structure in 2000, I discovered that the water level had dropped almost to 18 inches in a couple of days. I probably wouldn't have known this had I not crashed through the ice, but from that perspective, I could tell there had definitely been drainage. Returning that spring, I walked the dam to look for holes and found several muskrat burrows, but none of them were draining the wetland. For the next 2 years the wetland continued to drop soon after a

Beaver-dug channel which breached a dam, causing the failure of an 11-year-old constructed emergent wetland in Rowan County in the Daniel Boone National Forest in Kentucky.

heavy rain, and periodic inspections failed to find the problem.

After checking the water-control structure for leaks and walking the dam again in the spring of 2003, I rested for a moment of reflection, only to hear the faint trickle of water. Searching for the source of the noise, I found a weak stream of water emerging from the base of a willow growing along the backside of the dam. I entered the wetland along the inside of the dam and walked back and forth, going ever deeper until a series of holes were felt more than 3 feet below the surface. Stomping around the holes to stir up sediment, I went back to watch for water emergence. A few minutes later, muddy water seeped out by the willow tree, proving that the deep muskrat holes were draining the wetland. I then removed the sliding gate from the water-control structure to drain the wetland before leaving the area. Returning with a backhoe later that summer, I found two muskrat holes at the base of the dam that went through the entire dam. The backhoe filled and packed the holes, and the wetland is now performing as designed. What I learned from this experience is that it would have been more efficient to drain the wetland the first year the problem was detected so that a thorough visual inspection could have been made of the dam below the water level.

Beaver can make a large hole or channel through a dam, resulting in immediate failure of a wetland and a considerable amount of erosion. Finding a beaver-dug hole is seldom a problem. Filling them requires considerably more soil than does filling a muskrat burrow, often requiring that a backhoe operator borrow soil from a nearby hillside.

Often wetland dams have been neglected and have grown up to trees and shrubs, many of which are covered with thorns, making it difficult, if not impossible, to thoroughly inspect a dam for leaks. Mowing these dams with a bush hog and tractor will again make it feasible to inspect the dam. Should trees have grown too large to mow, a dozer can be used to clear the dam of woody vegetation. If this is done, ask the dozer operator to look for holes and repair them during the clearing operation.

Tree Roots in the Dam

Trees growing on a dam may eventually cause a wetland to fail, especially if the dam is narrow. Roots loosen the soil and provide small channels for water movement, and when growing over the length of a dam they can have a large impact on the hydrology of a site. Finding large trees growing on the dam of an older wetland that fails to hold water during the summer often indicates that roots are contributing to the failure. Smaller trees and shrubs can be effectively removed from a dam by a tractor and bush hog, whereas larger trees are more efficiently removed by a dozer.

Repairing a wetland is similar to fixing a car—work on the obvious things first before moving on to less common and more costly needs. Leaky water-control structures and holes in the dam are obvious reasons for wetland failure. Having exhausted these possible reasons for failure, you should be ready to move on to the more uncommon and obscure reasons.

Determining How the Wetland Was Designed to Hold Water

Wetlands are generally formed by trapping surface runoff with impermeable layers of soil or by exposing an elevated water table. Knowing how the failed wetland was planned to hold water can provide clues to its repair.

Test the soils in the bottom of the wetland and dam to help identify how it was designed to hold water. Soils that are high in clay and silt were probably used to trap runoff. Soils with sand, gravel, and rock would hold water only if the water table was near the surface or if a synthetic liner was used.

The main reason wetlands constructed of silt or clay soils fail is that water escapes under the dam. Water can leak under a dam via a buried topsoil layer, drainage tile, or crayfish burrow; along tree roots or poorly packed soils; or through gravel and sand layers. Thwart water from going under a dam, and a wetland should function as planned.

Using a backhoe or excavator to dig test holes in front of the dam can help identify how water is leaving a wetland. Test holes will help determine if a topsoil layer was left under the dam. Since topsoil is porous, a layer left under the dam can provide a path for water to leave the wetland by traveling under the dam.

Digging test holes just in front of the dam along its inside slope can help reveal the presence of deeper gravel and sand layers that may be carrying water under the dam. Subsurface permeable layers of gravel can be 2 to 10 feet thick and often rest on top of bedrock or a layer of clay. Digging test holes at 100-foot intervals along the inside of the dam will usually reveal the presence of buried permeable layers that are causing the wetland to leak.

Topsoil beneath the Dam

Randy and Tammy Cox dreamed of building a small cabin that would overlook a restful pond, a place where they could catch fish right off the deck. Their home in Pomeroyton, Kentucky, within the Daniel Boone National Forest is near the top of a mountain. They were told that the small dry pond they owned above their house had been built nearly 50 years ago. Tammy said, "The pond used to hold a little water in the spring, but about 15 years ago, it just dried up." Osborne Dozer Service had worked for Randy's family before, so they hired Billy Osborne and his 1150E Case dozer to repair the pond.[4]

Billy found that the old pond was basically a patch of dry forest when he showed up. Tulip poplar and sycamore trees over 20 inches in diameter were growing on the dam and down in the bottom. Tammy wanted most of the trees left, but Billy told Randy they would have to be removed in order to complete the repair. Randy gave Billy the nod to proceed with what he believed had to be done.

Billy's first step was to remove the trees growing on the dam and from within the pond. He found silt clay soils in the dam, which indicated the site should have held water. He then scooped out the 3 inches of topsoil that had formed in the bottom of the pond over the years.

Billy began digging into the inside toe slope of the dam and discovered a 6-inch-thick layer of topsoil that went completely under the dam. The layer was still dark and full of tree roots, and it was acting like a sponge to move water beneath the dam. After finding the topsoil layer under the dam, Billy pushed the existing dam into the middle of the pond, exposing the topsoil layer for its removal. The next step involved digging down into the soil beneath where the new dam would be built to sever tree roots and to locate permeable layers. He only found hard red clay beneath the dam's location, which made for an excellent core.

Billy ended up making the pond about 0.66 acre in size, which was three times larger than it was before, and about 12 feet deep. The entire rebuild took about 25 hours of dozer work that included removing trees from about 0.5 acre of hillside to make the field larger. He remembers billing the Coxes for $2,100.

Pond owned by Randy and Tammy Cox in Menifee County, Kentucky, filling with water for the first time in 50 years after a layer of topsoil was removed from under the dam.

I visited the pond only 2 months after it was repaired to find over 6 feet of water in it already. Tammy and Randy had stocked it with fish and built their cabin with a deck over the water's edge. Tammy said, "We spent a whole lot more than we estimated for the pond. The work seemed to go real slow. I looked after another day of work and said I could have moved that much dirt with a wheelbarrow." Tammy was upset that they could not save the trees and was still not convinced they had to be removed. The pond was obviously gaining water, showing that it is possible to repair a pond or permanent water wetland 50 years later.

Gravel beneath the Dam

The Forest Service received funding from Ducks Unlimited, Inc. to establish sixty emergent wetlands for the Wild Wings Project on the Daniel Boone National Forest in 1991. Twelve of the shallow marshes were built near Scott Creek in Rowan County by establishing low-level dams to keep runoff from reaching the creek. When inspected in the spring of 1992, six of the Scott

Creek wetlands failed to hold water. This disturbed me greatly, as I was on the site during construction and was sure that the wetlands would succeed. Wild Wings was our first cooperative project with Ducks Unlimited, and the reality of failure not only was embarrassing but also could jeopardize future restoration opportunities. Reviewing the details of construction, I knew that we did not leave topsoil under the dam, nor did we place it in the dam. We did prepare a shallow core for each dam by using the corner of the dozer blade to create a 3-foot-deep furrow, packing silt loam soil in the trench to finish the core.

I walked the bottom of each failed wetland that spring and found numerous mud chimneys made by burrowing crayfish. Looking down into their burrows, I could not see water in the deep holes. I returned to the office to retrieve a 5-gallon bucket, just to see if I could fill the holes with water. Some of the holes would take three bucketsful, while others could not be filled. Thinking that the crayfish were to blame, I searched the then-new Internet for a solution and returned the next day with 20 gallons of bleach, pouring some down every hole I could locate. I finally used a shovel to plug each burrow with soil. Two weeks later, even after 3 inches of rain, the crayfish had reopened their burrows and the wetlands remained dry.

That wet spring gave me plenty of time to think and to ask others about possible causes for our wetland malfunction. I visited the six wetlands with contractors, biologists, and representatives from the NRCS in an attempt to discover what went wrong. I walked the area between the creek and the new dams after heavy rains to look for missed drain tile outlets and even searched the bottom of each wetland to see if water was leaking down into cracks or collapsed drain lines. After much contemplation and advice, I decided to line two of the failed wetlands with a layer of clay.

In the summer of 1992, I contracted with dozer operator Dick White to push the organic material from the bottom of one of the failed wetlands over to the edge of the wetland. White pushed

clay from the adjacent hill down into the wetland and packed it in thin layers up to 12 inches thick, then covered the clay with the saved organic material. I did the same with the second failed wetland nearby, only John D. Smith used a backhoe to move the clay from a more distant borrow source over into the wetland. Rains that came in the winter and spring of 1993 showed that the clay liners had failed. Crayfish had tunneled through the clay, and the wetlands remained dry.

After even more discussions and study, I brought back Smith and his backhoe in the summer of 1994 to see what the soils were like deep down in three of the failed wetlands. We started digging holes inside each failed wetland at the base of the dam and found a 3 to 4-foot-thick layer of silt loam on top of 5 to 7 feet of gravel, with a bedrock layer below. Water was moving through the gravel but over the surface of the bedrock. We found that the silt loam layer became thicker and the gravel layer thinner with increased elevation and distance from the creek. Near the base of the hill far above the creek, we found 10 or more feet of silt loam on top of only 1 foot of gravel, and the gravel again rested on bedrock. The test holes showed that water was traveling down the crayfish burrows into the gravel layer, then under the dam to the creek. The gravel on top of the bedrock was acting like a large underground stream that had to be interrupted for the wetlands to work.

We brought in a small dozer and a backhoe in the fall of 1994 in an attempt to repair two of the six failed wetlands, deciding to work on those not lined with clay. The dozer scraped the topsoil and organic material from the dam and from a 50-foot-wide strip in front of the dam. The dozer made a level bench on the inside slope of the dam for the backhoe to set up and begin work. Starting at the base of the hill where the dam began, the backhoe dug a trench down to bedrock along the entire inside of the dam. The silt loam we removed from the trench was piled inside the wetland, and the gravel was placed along the backside of the dam. The dozer moved the silt loam from in front of the dam over into the trench and then attempted to pack it with one track. The dozer restored the slopes on the dam and covered exposed surfaces with the saved topsoil.

Unfortunately, our inspection conducted in the spring of 1995 showed that the backhoe treatment did not work. Portions of the narrow core trench had settled up to 4 feet, leaving a deep open trench in the dam. Walking on the dam was an unnerving experience, as one would suddenly break through a crust that had formed a bridge over the surface of the narrow core trench. It appeared that water continued to move under the dam, through the loose and poorly packed soil placed in the narrow core.

Discouraged but not beaten, we decided to try yet another technique during 1996 in an attempt to repair the largest of the failed wetlands at Scott Creek, this one being 3 acres in size. This time, we brought in two large dozers and an excavator. The project began with the dozers scraping off the topsoil from the dam and from a zone 150 feet in front of the dam. The dozer then cut into the inside slope of the dam to make a level bench from which the excavator could operate. The excavator dug a 12-foot-wide trench, down to bedrock, along the entire inside of the dam. Beginning at the base of the hill where the dam began, the excavator removed the silt loam from near the surface and piled it in a row inside the wetland. The gravel was then removed from the trench, down to the bedrock, and placed in a row along the backside of the dam. The excavator operator made sure that loose plates of rock on top of the bedrock were also removed. During the coring operation, slow-moving streams of water could be seen flowing on top of the bedrock.

Once the excavator had a 100-foot head start, the dozer began pushing silt loam from in front of the dam over into the core trench. A path was made down into the core trench, so the dozer could spread, pack, and build up layers of silt loam in the core along its entire length. There were times when the dozer was only 10 feet from the excavator bucket, pushing in silt loam as the gravel layers caved in because they were so high

(top) First wetland repaired at Scott Creek by using an excavator and dozers to dig and pack a core down to bedrock beneath the already constructed dam.

(bottom) Last wetland repaired at Scott Creek by using an excavator and dozer.

and loose. The entire core was packed with silt loam soil up to the original ground surface, providing a solid foundation for the new dam.

We proceeded to build the new dam on top of the core. The rebuild gave us the opportunity to place a more gradual slope on the inside of the dam, thus allowing for a greater diversity of plants. The great quantity of gravel we removed from the core was gently sloped to form the back-side of the dam, which helped make the wetland look more natural, as it was now difficult to see the dam's profile. A layer of topsoil was placed over the dam and the bottom of the wetland to complete the job.

With apprehension, we visited the wetland in the spring of 1997 to find that it held water. It now looks and functions as designed, and we have used the technique to repair all of the failed wetlands at Scott Creek. During the winter, we now watch bald eagles and osprey hunting over the wetlands. In the spring, it is common to flush hundreds of mallards, ring-neck ducks, and blue-winged teal from the waters. During the summer, Canada geese, mallards, hooded mergansers, and pied-billed grebes use the wetlands for nesting, making it hard to believe that they were a rare sight in eastern Kentucky only a few years ago.

Based on the Scott Creek experience, I have learned that it is essential to have qualified individuals who have a sincere interest in and knowledge of wetland work present at a construction site at all times the contractor is working to successfully complete a reconstruction project. Each day situations arise at the job site that must be discussed with the contractor to ensure success, including questions about how deep to make a core and where the dam should end. Leaving a contractor unattended provides an environment where temptations are there to take shortcuts that later show up as failures when the system does not hold water as planned, and with no reasonable way to find out what went wrong without redoing the entire project. Under strong direction from management, I have left contractors unattended for several days only to find later that dams were built too steep, depths excavated too great, and heaven only knows what was placed in the core.

Missed Drain Lines

It is generally not reasonable to test a failed wetland for the presence of drain tiles, tree roots, or crayfish burrows that may be carrying water under a dam. So much digging would need to be done to make such a determination that it is less expensive to assume their presence and rebuild the wetland accordingly. An indication that drain lines were missed during construction is that water levels dropped rapidly after heavy rains. Most drainage systems are designed to lower standing water levels in fields by at least 12 inches a day. Water levels in a wetland built on top of an active tile system can be expected to act the same way.

One may have an idea that drain lines, crayfish burrows, or tree roots were left under a dam by interviewing people who designed and built the wetland. When doing this, pay close attention to how the coring operation was completed. Most contractors do not understand the importance of coring and barely scratch the surface of the soil on which they build the dam.

Drain lines can be buried deep beneath wetland projects, as evidenced by this advice from Henry French in 1903: "Three-foot drains will produce striking results on almost any wet lands, but four-foot drains will be more secure and durable." He then urged drainers to bury clay tiles well below the reach of subsoil plows.[5] Clay drain tiles buried 8 feet deep in the early 1700s were found to be in excellent condition when uncovered near Lincolnshire, England, in 1831.[6]

A ground-penetrating radar unit (GPR) can help identify where underground features such as drain lines are located in a failed wetland. The GPR will show where something is buried in the ground, but it does not identify what the object might be. Should you have one available for your use, I suggest pulling the GPR across the bottom of a failed wetland in a series of lines, spaced 30 feet apart, which are parallel to the dam. A brightly colored plastic flag should be placed over the location where each item is detected in the ground. Upon completion of the test, the presence and location of buried drain lines can then be suspected by rows of flags that pass downhill to the dam.

Roland Riparian Restoration Project

Dennis Eger began building wetlands in 1977 and now works as an assistant manager for Public Access South and as the forest wildlife manager for the Division of Fish and Wildlife with the Indiana Department of Natural Resources. During the summer of 2000, the Indiana DNR began restoring a series of four shallow marsh wetlands adjacent to an existing wetland as part of the Roland Riparian Restoration Project. The project was designed to inundate a total of 68 acres in Martin County, Indiana. Construction involved using dozers to restore natural levees in old fields along the Lost River. Workers cut through a number of buried tile lines during construction of the wetland cores, and after blocking and filling the tiles, they built dams on top of the core from silt-clay-textured soil.[7]

The spring after construction, workers opened the water-control structures on two of the new wetlands so that moist soil plants would germinate and grow. Completing this project required the cooperation and funding of a number of groups and organizations, so a dedication was scheduled that fall to help celebrate its accomplishments. To prepare for the dedication, the water-control structures were closed so that the wetlands would fill from precipitation. Unfortunately, one of the wetlands did not gain water, even after heavy rains. With the dedication looming near, the Forest Wildlife crew borrowed an 8-inch-diameter trash pump to fill the wetlands directly from the river. Regrettably, water levels would not build in the wetland even with pumping and failed to spread out over the moist soil plants blanketing the bottom. Having been involved with the actual construction of the project, Dennis was confident that the wetland could be made to hold water.

Suspecting a missed drain tile, the Forest Wildlife crew walked the banks of the river between the newly constructed dam and the river channel. While walking along the river, one of the crew heard water flowing and called Dennis over to investigate. They eventually found water gush-ing from the end of an 8-inch-diameter clay tile, about 20 feet from the river. The clay tiles looked like they were from the early 1900s, but there was no record of them having been placed in the field. The crew marked the location of the outlet with a brightly colored ribbon so that they could find the tile line with a backhoe. Returning with the backhoe, they had trouble finding the tile in the new wetland and eventually unearthed it at a depth of 4 feet in the bottom of what had been a shallow drainage ditch in the field. The main tile line had started at the lowest point in the wetland, following a low area until it then passed under the dam at an angle to the river. The inside of the tile was washed clean by the sheer volume of water it had been carrying. The backhoe severed the tile line in two places, one on the inside and one on the outside of the dam. They placed concrete in the end of the tile within the wetland and packed both holes with clay. The wetlands looked great for the dedication and are now holding water as planned.

Dennis had the Forest Wildlife crew examine each new wetland construction site for active tile outlets by walking river- and streambanks after a heavy rain. Most often, water was flowing out of the end of the drain tiles and could be heard. A pool of muddy water in the river also alerted them to the presence of a tile outlet, as sediment entering a tile line from within a new wetland can be carried directly to a river via a buried tile line. Tile outlets are marked with brightly colored ribbons placed on vegetation that can easily be seen by an equipment operator. The ribbons tell the operator either to core deeper with a dozer to cut a tile or to return with a backhoe or excavator to cut and plug the tile. While restoring wetlands in Indiana, Dennis has found 4-inch-, 6-inch-, 8-inch-, and even 12-inch-diameter clay tiles that were used to drain low-lying fields.

There are risks to using a narrow trench dug by a backhoe to find and plug drain tiles, primarily because it is difficult to pack soil back into the narrow trench, and water pressure from an otherwise intact drain line can bypass such an obstruc-

(top) A buried clay tile line that carried water into the Lost River since the early 1900s and continued to function even after a series of wetlands were restored on top of the line within the Hoosier National Forest in Indiana. While searching for the cause of the leak, the Forest Wildlife crew heard water running from the tile outlet along the riverbank; after locating the outlet, they removed a section of tiles and blocked their path with a backhoe. (Photograph by Dennis Eger, Indiana Department of Natural Resources)

(bottom) View of Cell Four of the Roland Wetland Project, demonstrating that persistence is often rewarded when you keep working to identify how water is leaving a failed wetland. (Photograph by Dennis Eger, Indiana Department of Natural Resources)

tion. The trench bisecting a drain line should be wide enough for a dozer to enter for packing, as water can continue to leak under a dam if soils remain loose in the trench.

The importance of digging and packing a deep core beneath the dam location or along the lower edge of a wetland project cannot be overemphasized. As described earlier, many techniques that were used to move water beneath the surface of the ground are still working today.

Ballard County

Pat Hahs and Charlie Wilkins, wildlife biologists with the Kentucky Department of Fish and Wildlife Resources, were flying a waterfowl survey in western Kentucky during the spring of 2003 when they spotted a brown muddy field where a wetland had been the day before. Circling over what had been Cook Slough, they observed a long brown plume of soil that originated in Humphrey Creek that flowed into the Ohio River and then extended downstream for miles to the Mississippi River. The 30-acre wetland had been constructed in the late 1960s within the Ballard County Wildlife Management Area. Ground inspection revealed a huge, 150-foot-wide triangle-shaped crater that was 14 feet deep within what used to be the wetland. The washout had consumed a section of the 8-foot-high dam, the flashboard riser water-control structure that had been partially filled with concrete, and a 30-inch-diameter, 40-foot-long corrugated steel drainpipe. Pat said, "It was like a landslide on flat ground. We'd never seen anything like it, kind of an eighth wonder of the world to us." Peering into the canyon, they could see a clay drain tile about 4.5 feet below the original ground surface.[8]

Charlie says no one was sure what caused the washout. The creek was low, and there was no flooding: "We were pumping water at the time and it was within 6 inches of the top of the dam. Beavers were common in the area, and it is possible they had made a slide over the top of the dam, and it is also possible that water may have flowed over the dam via the beaver slide."

After talking with Charlie, I believe that it was also possible that a main line for an old drainage system was buried at the location and that the washout was caused by water within the wetland finding a direct route into the drain line. The entire wetland had apparently been built on the same location as a natural wetland that had been drained years ago with clay tile. The dam for Cook Slough may have been built over the main tile line leading to the outlet, which would have been buried deep in order to drain the natural wetland. Old plugged drain lines can suddenly open under the pressure of deep waters created on the surface.

Charlie knows that clay tile was used to drain much of the land on the Ballard County Wildlife Management Area. Over the years, he had noticed wet spots along the banks of Humphrey Creek during dry weather, which he always believed indicated locations for drain tile outlets. "We didn't find drain lines when we replaced the water-control structure and rebuilt the dam 3 years ago, but we may not have looked deep enough," he said. What makes me believe that portions of the drainage system were still working within the wetland was that it dried out enough to plant soybeans or corn each spring after the water-control structure was opened.

A new dam has been built up from the blowout location to bring Cook Slough back, and, fortunately, no buried drain lines were found when coring the new dam. The new dam is the same elevation as the old dam, but the wetland is now 4 acres smaller in size.

Minor Clark Fish Hatchery

The dry 6-acre pond was one of 100 built for the Minor Clark Fish Hatchery in 1972 by the Kentucky Department of Fish and Wildlife Resources. Unfortunately, this, the largest of the constructed ponds, happened to be located next

Large washout, where a 30-acre constructed wetland disappeared overnight, near Humphrey Creek within the Ballard County Wildlife Management Area in Kentucky. (Kentucky Department of Fish and Wildlife Resources photograph)

to the main highway leading to the popular Cave Run Lake. Al Surmont, fisheries biologist, said, "We ran an 8-inch water line into the pond until it was 2 feet deep, only to find it would disappear overnight."[9] They added potassium permanganate dye to the water after one of the pumping sessions in an attempt to find out where all the water was going. Fortunately, they found colored water leaking from the backside of the dam. Using a backhoe to dig down along the back of the dam just above where the colored water emerged, they found a buried line of clay tiles that were 4 inches in diameter and octagon shaped, carrying water from the pond. They cut out a section of the line with the backhoe and filled the bottom of the hole with bentonite clay, packing silt loam soil in to fill the trench. The pond has held water since then and looks beautiful from Highway 801 with mountains in the background.

Be aware that more than one layer of drainage lines may be present on a wetland construction site. Mark Lindflott of the NRCS found three separate drain line systems that had been buried at different times and at different elevations while

Six-acre constructed pond at the Minor Clark Fish Hatchery in Kentucky. The pond failed to hold water until a line of buried clay drain tiles that passed under the dam was discovered and cut, and the section was blocked with clay and silt loam soils.

working to restore a 30-acre wetland in Iowa. A number of these drain lines were still working when he constructed the wetland.[10] The same problem was reported from Wisconsin where some sites were found to contain several layers of tile due to generations of farmers installing drainage systems in the same location.[11]

Clay Tiles Open Years Later

John Meredith and George McClure have developed great respect for each other by working together for many years to complete drainage projects in Ohio. As a district conservationist for the NRCS, John had inspected many jobs that drainage contractor George had completed. After seeing the quality work he finished, it made sense to John to hire George to build a wetland on the farm he had just purchased. John was able to obtain enough financial assistance from the Ohio Department of Natural Resources to cover about 50 percent of the cost of the project. So in 1997, he hired George to build the 3.5-acre emergent wetland with his dozer near Little Salt Creek

in Jackson County. They chose to build the wetland in a low area with hydric and mottled silt clay soils, located between an open ditch and the creek. The field had last been used to grow corn and soybeans in 1990.[12]

To build the wetland, they shaped a long, natural-appearing dam with gradual 15:1 slopes designed to impound water to a maximum depth of 3.5 feet. They protected a large pin oak tree growing in the middle of the field from drowning by keeping water away from its roots. John installed a 6-inch-diameter PVC drain pipe in the dam with a plastic water-control structure over its inlet. George found a few clay tiles during construction but did not pay much attention to them as they were plugged with soil.

Unfortunately, upon completion, the wetland failed to hold water. Situations such as this can be especially embarrassing to people in their positions. John did some investigating after heavy rains and eventually found water coming out of the ground in a ditch about 200 feet below the wetland. He knew then that buried drain lines were the problem and used a tile probe to locate two lines below the inside toe of the dam. He used a shovel to dig 4 feet deep to expose the lines, blocking them with soil and bentonite clay. Still, the wetland would not work. Two years after he built it, George returned with a backhoe to dig a trench along the inside of the dam, finding seven 4-inch-diameter clay drain lines that were spaced 50 feet apart and working as well as the day they were installed. George cut through these tiles and packed the trench with silt clay soil, successfully repairing the wetland. Both John and George claim that water entering the wetland had flushed open the plugged underground tiles. "I guess they're never really blocked," said John.

As we walked around the wetland one late January afternoon, I was struck by the beauty of the lone pin oak reflecting in the water. John described how the wetland is home to muskrats, as evidenced by their many feed piles, along with ducks, geese, frogs, and many aquatic plants. We pondered over why every one of the cypress trees he planted had been rubbed by the deer. I left encouraged by how these partners in drainage were now working together to return wetlands to the landscape.

Rock Channels beneath the Dam

Back in the 1960s, the USDA Forest Service received funding to build a number of small, permanent water ponds on mountain ridges for wildlife habitat improvement within the Daniel Boone National Forest. Several of the ponds constructed in Bath County never held water. Forestry technician Richard Hunter and dozer operator Dale Darrell worked to repair two of these dry ponds in the summer of 2003. They first cleared trees that were growing on the dams and in the bottom of the failed ponds. Next, they dug into the dams and found they were made of clay.[13]

The first site was located at the base of a limestone cliff. There they dug down into the pond at the base of the dam and began pushing out piles of clay. Around 4 feet below the surface, they hit a large limestone rock that was surrounded by a layer of sand and gravel. They removed this rock and mixed the sand and gravel with clay, packing the mixture back into the core and in the bottom of the pond. The small wetland is now holding water as designed. Dale believes that the original contractor most likely found the large rock in the bottom years ago but could not remove it with the straight blade on his dozer. Dale was able to extract the boulder by getting under it with the modern seven-way blade on his dozer, thus blocking the path by which water left the wetland.

When digging into the bottom of the second pond, they uncovered limestone rock 3 feet below the surface. Further digging showed that the rock formed a layer beneath the entire pond, and cracks were found in the rock. The dozer rooted around the bottom of the pond, moving, mixing, and packing clay over the rocky base. Dale then moved soil from an area immediately upslope of the pond to cover the bottom of the pond with 12 inches of packed clay. Originally designed to hold water year-round, the pond was changed into an

George McClure *(left)* and John Meredith *(right)* in 2005 at a wetland they restored in 1997, on John's farm in Jackson County, Ohio.

ephemeral wetland. The treatment worked, and the wetland is now functioning as designed.

Adding More Water

It may be possible to lengthen the time that water remains in a wetland by increasing the quantity

Repaired small wetland in Bath County, Kentucky, that failed to hold water for over 40 years until it was restored by removing a large rock surrounded by porous soils from the bottom of the site.

An initially failed 0.5-acre wetland, located near Beaver Creek in the Daniel Boone National Forest, built from sandy silt loam soil. The overflow from a 5-acre wetland above it was then directed into it, causing it to hold water.

of water allowed to enter the system. The soil in some wetlands is porous enough that water seeps into the ground faster than it is replaced by rainfall. Look for roads constructed years ago that are uphill from the wetland site to see if ditches divert water away from the wetland. Installing several culverts in a road uphill from a restored wetland can add a significant amount of water to a wetland, sometimes replacing the historic watershed. Small streams that go around a restored wetland can usually be made to enter a wetland. A close examination of these small streams may show that they were channeled and straightened years ago and that they historically fed the wetland. A small stream that enters a larger stream at a right angle is a sure sign that the original stream was channeled and moved over to create more farmland.

Limestone Has Its Problems

Thad and Amy Ross were determined to repair the 0.25-acre pond they found hidden in the woods on the 5 acres they purchased near Leesburg in Central Kentucky. They had left the big city of Lexington to move into their new country home in the rolling hills and limestone-dominated topography of Harrison County. "We concentrated on getting the house together the first year after we moved in," Thad said, "but then I couldn't help but start working on the dry pond. When I began working on it, it was nothing but a glorified mud hole with a great big dam and a little puddle of water at one end. It was so dry that I could mow the bottom."[14]

Thad examined the pond on a regular basis to see if he could determine what was happening to the water. After heavy rain, water entered the pond from a small stream. He said, "It would drain out almost as fast as it came in; you could watch a whirlpool above the leak that would last until the water was gone." The water disappeared down a 5-inch-diameter hole in limestone rock in the bottom of the pond near its upper edge. Thad found that the water emerged 200 feet downstream in a small creek below the pond. He then realized that the pond had been built on top of a small limestone cave.

One way he tried to repair the pond was by pouring several bags of dry concrete mix into the hole during the summer of 2000. The action appeared to plug the leak, as the pond gained water all summer. Unfortunately, heavier rains formed a channel around the cement and drained water down into the hole, again drying out the pond.

Thad searched the Internet to see if he could discover how to repair the pond and came across suggestions for using bentonite clay to fix leaks. Encouraged by the possibility, Thad dug out the soil from around the hole in the rock and, while water was flowing, poured in bentonite. After pouring five bags of bentonite into the hole, the leak was plugged. Over the next couple of weeks, the pond filled with 18 inches of water and then suddenly failed, flushing the bentonite into the stream. He tried this technique once again with five more bags of bentonite and experienced the same unfortunate results.

As a professional landscape architect, Thad knew when it was time to seek the advice of others, so he phoned the Kentucky Department of

Fish and Wildlife Resources, who referred him to the NRCS for help. The NRCS district conservationist met with Thad to look at the pond and recommended that he consider filling the cavern below with mortar to stop the leak.

Thad found out that a local contractor, Richard Williams, had experience with the mortar technique, using it to successfully repair ponds at horse farms in the area, so in the summer of 2003, he hired him for the job. Richard drilled a series of holes about 10 feet apart, in a straight line across the lower end of the pond, perpendicular to what he believed would be the cave channel. He used a 3-inch-diameter masonry bit to drill through the rock into the limestone cavity. Using a portable concrete pump, he injected liquid mortar into the cavern. By chance, a small puddle of water remained in the pond that could be used to mix the mortar.

Richard charged Thad for each bag of mortar he pumped into the hole. Richard checked with Thad on whether or not he should continue work after each pallet of forty bags was used. Working on the project from time to time over a 2-week period, Thad eventually told him to "do whatever it takes to get it fixed," so he stopped checking until the job was completed, using a total of 240 bags of mortar. The pond filled completely over the winter, staying full all spring, and then dropped only about 8 inches in the dry summer weather. Thad was pleased to flush black ducks and wood ducks from the water the first winter it stayed filled.

Thad led me down to the pond a year after it was repaired. Green frogs and bullfrogs jumped into the water, and a green heron squawked as it flew away. Dragonflies patrolled the surface of the water as gray tree frogs called from the woods. Looking out over the pond, Thad proudly stated, "It's the best $3,000 I ever spent."

Michael Jackson Situation

Brothers Anthony and Sonny Utterback want their customers to be satisfied with the work they do in eastern Kentucky. Their reputation for excellent work and reasonable rates were the main factors that influenced Michael Jackson to hire them to build a road and large pond on his farm in Mason County. Construction of the road involved moving a considerable amount of soil across a hollow, enough so that the Utterbacks recommended Jackson use the fill to construct a smaller pond, about 0.25 acre in size, which would be clearly visible from the new home he was building that fall.[15]

While looking at the job, the Utterbacks found a small sinkhole in the bottom of the small pond they had received permission to construct. The sinkhole was knee deep and full of grass, and it looked like someone had thrown a roll of fencing into it. During construction, they used their dozers to remove topsoil and debris from the sinkhole and replaced it with a thick layer of clay. Knowing the need for a good core, they prepared the base of the dam down to a layer of dense clay, then packed clay in layers on top of the core to create the dam. The road turned out well, and both ponds filled with water. Everyone was extremely pleased with the job.

One night the following spring, Sonny received a call from Jackson that the water in the

Thad Ross at the pond he repaired with mortar after it failed to hold water for 14 years, in Harrison County, Kentucky. Note the stressed sycamore that had been growing in the dry depression.

small pond had disappeared over a 4-hour period. When Sonny went out to investigate, he found a washed hole, about 12 inches in diameter, in the middle of the now-drained pond. The water from the pond was still emerging below the backside of the dam near the base of a large hickory tree.

The Utterbacks brought in a 580 Case backhoe to repair the pond. They dug a deep trench along the base of the inside toe of the dam and cut through a stovepipe-sized channel about 4 feet deeper than the original core made for the dam. One of them stuck his arm into the channel and found it to be lined with rock as sharp as glass, washed clean by the flowing water. They packed the channel with clay and rebuilt the inside toe of the dam. Over time, the pond never filled with more than 1 foot of water.

Sonny and Anthony returned to the dry pond the next time with an excavator and a dozer. They cut a bench into the inside toe slope of the dam and used the excavator to dig a 12-foot-wide trench deep into the ground along the inside edge of the dam. They found and cut through the limestone channel with the excavator, then cleaned out the channel with a shovel. The limestone channel had apparently originated in the sinkhole and paralleled the original ground surface but was 6 feet below the surface as it went downhill. As soil was used from the bottom of the pond to build the dam, the thickness of clay left over the channel was reduced. Water pressure and seepage from inside the pond found its way into the channel and under the dam similar to a buried drain line. They packed five bags of bentonite and five bags of concrete in the channel. They also packed layers of clay into the core and again rebuilt the dam. Fortunately, the pond filled and has held water ever since.

Sonny told me that these repairs cost him and his brother several thousand dollars, but they had no choice but to keep coming back as "our name was on the thing." Karst topography is common in the Fleming and Mason County area where the Utterbacks live, and they both know of a number of ponds constructed over limestone layers by other contractors that fail to hold water.

Where Was the Water Table When the Wetland Was Built?

When rocky or sandy soils are found in a failed wetland, the obvious conclusion is that the site was originally designed to intercept the water table. Do not automatically write off failed wetlands constructed on sites dominated by porous soils until they have been watched for a couple of years. Water tables can fluctuate seasonally, and drought conditions may be responsible for the lack of water. It would also be wise to find out if the wetland ever held water as planned; if it did, there may be a few more options available for its repair.

Examining the bottom of a failed wetland designed to expose the water table can provide clues to its failure. Finding an abundance of aquatic plants is a good thing, as they indicate the presence of saturated soils and possibly a temporary drop in the elevation of the water table. Dig some test holes in the bottom of the wetland with a long soil auger to see if water rises in the hole. Water appearing near the surface could indicate a leaky water-control structure, a muskrat hole in the dam, or water passing under a dam through a buried drain line. If water does not enter the test hole, the water table may never have been high enough to form a wetland. A lack of hydric plants growing in the wetland also indicates a depressed water table and a location that never held water. Digging deeper holes with a backhoe is a faster way to find where the water table is located at a given time. However, setting up several PVC monitoring wells is the best way to tell what the water table is doing on a year-round basis. Wetlands on sites where the water table has dropped can sometimes be repaired by excavating deeper until waters are again exposed.

When Patience Pays Off

In the year 2000, Dennis Eger, Forest Wildlife group leader for the Indiana DNR, worked on constructing a 60-acre wetland called Provence

Pond in Henry County near Newcastle, Indiana. Dozers and a scraper were used to build 4,800 feet of levee to a maximum height of 12 feet. Workers dug the core for the dam down 6 feet and then built a dam of clay and silt clay on top of the core, cutting and plugging all buried drain lines. They noticed a large number of crayfish burrows during construction, many of which had diameters of 2.5 inches. Dennis was surprised to find so many active crayfish burrows, considering the site had been cultivated since the early 1800s. The wetland failed to hold water after construction, resulting in great disappointment to all those involved.[16]

Not willing to accept failure, in 2001, Dennis decided to dig deep test holes on the construction site with a large excavator that had a 30-foot reach. He found that the wetland was built on an area with a 2- to 3-foot-thick cap of silt clay that overlaid a number of alternating layers of sand, gravel, silt, and clay extending down 30 feet or more. The most unusual part was that these layers were vertically, not horizontally, arranged. Provence Pond had been built on an area that had been glaciated, helping to explain the presence of so many layers and their odd orientation. Some of the sand layers were full of water, whereas others were dry. The large crayfish burrows extended down into the water-bearing layers, and one green-colored crayfish was even observed at a depth of 30 feet. They were unable to identify a technique that could keep water from traveling down the crayfish holes, into the permeable layers, and under the dam.

I had originally included this story to illustrate that some wetlands can't be repaired at a reasonable price, when Dennis sent an e-mail to me with photographs showing Provence Pond filled to capacity in the spring of 2004. His theory for success was that since the wetland had such a small watershed, it must have taken years for rainfall to first saturate the ground and then fill the pool. He believes that the rising water levels may have made it unnecessary for crayfish to burrow and that, possibly, wave action worked to fill and plug some of the crayfish burrows with soil. We may never know all the reasons for this unexpected success story, but it does show how patience can help when faced with a decision to spend a great deal of money to fix a wetland that isn't functioning as planned.

Failed Wetlands That Must Stay That Way

In 1997, I built two wetlands near the mouth of Brushy Fork near Cave Run Lake in the Daniel Boone National Forest. I had used a handheld soil auger with a 5-foot-long handle to sample soils on the site before construction. The samples revealed silt loam soils down to the bottom of each hole. We used two dozers to move topsoil off the site, excavated a core down 4 feet deep beneath the dam locations, and built the dams from silt loam soils. One of the dozers even got stuck while building the small 0.5-acre wetland next to the lake. Water pumped up from crayfish holes near its dam and turned the site into a quagmire. The smaller wetland held water as planned and is believed to be fed by both surface water and groundwater. The larger 2-acre wetland failed and only holds a small pocket of water near the drainpipe.

Walking over the larger wetland a year later, I found a number of dry crayfish holes in the bottom and suspected that water was traveling under the dam. In 1999, I brought in John D. Smith and his backhoe to dig deep test holes on the site. We found that the area had a cap of around 5 feet of silt loam on top of 6 feet of gravel. The gravel was on top of the bedrock layer, with water running on top of the bedrock. Thinking that it might be possible to rebuild the site with a deep core constructed by an excavator, we began looking for sources of silt loam to replace all the gravel that would have to be removed. The wetland was bordered on two sides by Brushy Creek, on its lower edge by the smaller successful wetland, and on the upper side by a mountain slope containing gravelly soils. The only way to fill a deep core was to truck in silt loam at a considerable cost, and we found this to be impractical in light of current budgets.[17]

Constructing Wetlands to Last

It is entirely possible to build a wetland that will function for hundreds, if not thousands, of years. I have visited a number of small wetlands in Kentucky averaging 30 feet in diameter that were dug by hand and with the help of livestock from the late 1700s to early 1800s that continue to contain water, along with a diversity of plant and animal life. These ponds were dug with the help of hand-controlled steel scoops pulled by horses, mules, or oxen.[18] The soils on these early pond sites are generally fine-textured silts and clays. Some were built on top of springs, while others were compacted to hold rainfall and small amounts of runoff. Watersheds draining into these ponds are vegetated with trees or a mixture of grasses (or both), with little, if any, area of exposed soil.

Natural wetlands can tell us much about how to build a wetland that will last. Cliff Palace Pond is a small, natural ephemeral wetland that measures only 50 × 100 feet. Located on top of a sandstone mountain in Jackson County in the Daniel Boone National Forest, it is the last place one would expect to find water. The wetland is less than 2 feet deep, with a watershed only a little over 1 acre in size. The pond contains silty to sandy clay-textured soil and appears to hold surface water, not groundwater. Radiocarbon analysis of sediments reveals that the pond is about 9,500 years old, with pollen analysis showing that it existed as a cattail marsh for several thousand years before becoming an open pool of water with scattered buttonbush.[19] Cliff Palace is surrounded by a gentle 4 to 5 percent slope and lacks a defined inlet or outlet channel.

Bear Pond is another small natural wetland measuring only 15 feet in diameter on top of a narrow sandstone ridge in Powell County in the Daniel Boone National Forest. Its silt clay soils hold surface water year round, as indicated by the presence of spotted newts and buttonbush. Radiocarbon dating indicates an estimated age of 790 ± 40 years BP.[20] The pond is situated in a shallow basin with no defined inlet or outlet, and its surrounding banks have only a 5 percent slope.

A number of factors explain how a constructed wetland can last for generations with little or no maintenance. The most important revolves around constructing a wide dam with gradual slopes. Placing a 20:1 (5 percent) or gentler slope on the front and back of the dam minimizes potential damage by tunneling muskrat and beaver. Creating the top of the dam to be 12 feet or wider also helps reduce problems that may develop later if a tree topples over and its roots create a large gap in the dam.

Another factor that promotes longevity of a wetland is a small watershed. Providing the wetland is constructed of impermeable soils, or is on a site with a high water table, there is no need to direct great quantities of runoff into a wetland constructed east of the Mississippi River. Wetlands that lack a defined inlet stream are less likely to fill with sediments washed in over time.

An important factor to consider when building even a small wetland is to provide a path for water to leave the wetland without damaging the dam. Directing possible overflow around the dam so that it follows a wide path down a gradually sloped spillway on undisturbed ground will prevent a channel from forming at the outlet that can eventually drain the wetland. Such sites are also less likely to be dammed and channeled by beaver.

Working to rapidly establish vegetation on a newly constructed wetland can give a wetland an important head start for a long life. Taking action to immediately seed winter wheat on exposed slopes and mulching with straw will help keep soils from washing down into and filling the new wetland.

There are many failed wetlands in Canada and the United States that could be fixed by stopping the path by which water travels under or through the dam. These stories of successful repairs illustrate that a dry wetland does not have to be accepted as a permanent status. Just as draining wetlands often occurred in stages over many years, returning them to their original hydrological regime, native plants, and animals can also require more than one attempt.

CHAPTER EIGHTEEN

Finding Funding

Knowing how to identify drained wetlands and to plan for their restoration will not accomplish much unless financing is available to complete the job. Fortunately, many government and non-government organizations have funding available for wetland projects. The following information can help individuals develop a successful strategy for obtaining the funding needed to implement a wetland restoration program.

Preparing a short proposal is an important first step when looking for money. The proposal should contain objectives, locations, methods, budget, time frame, people and organizations involved, and whom to contact for more information. Try to write this proposal in four pages or less, and be ready to give it to prospective contacts.

Work to have a project ready to implement. Make sure all necessary cultural resource surveys, endangered species surveys, permits, and National Environmental Policy Act documentation (if required) are complete before requesting funds.

Ask people to get involved in your project, as efforts that involve partnerships with a number of organizations and volunteers are more likely to be funded. In 2004, I helped complete the Kentucky Pride Wetland Restoration Project. The project began with a $700 donation from the Rowan County Wildlife Club; grew with a $1,200 donation of PVC pipe from Waterflow Systems, Inc.; was increased with $2,000 of native wetland plants from the Kentucky Department of Fish and Wildlife Resources; then grew with $50,000 from Eastern Kentucky PRIDE, $25,000 from Ducks Unlimited, Inc., $10,000 from the Steele-Reese Foundation, and $20,000 from the USDA Forest Service Herp Conservation Grant Program. We hired contractors with heavy equipment to move soils and recruited volunteers from high schools, universities, Boy Scouts, Girl Scouts, and the Painted Hills Garden Club to help build 140 wetlands, plant aquatics, place woody debris, and erect waterfowl nest boxes within the Daniel Boone National Forest.

Believe strongly in the task. Know the purpose of the project, and be ready to explain it to others in simple terms. Your confidence and enthusiasm will be contagious.

Call or visit the government agency, conservation organization, or foundation before submitting a proposal, and talk with the grant contact person. Remember that funding is given from one individual to another. Even though you represent separate organizations, the development of a relationship between you and the grantor will greatly increase the chance of success.

Demonstrate that you can be trusted to complete the project. Be ready to list previous projects completed, and show how your organization can get things done. Scott Manley from Ducks Unlimited, Inc., and Jennifer Johnson from Eastern Kentucky PRIDE say that delays and unfinished projects are among the greatest problems they face when providing funds for projects that other organizations are expected to complete.[1]

Use the in-kind value that your time represents and the funding that your organization commits to leverage matching funds for the project. Many organizations such as Ducks Unlimited, Inc., have a goal of multiplying each dollar they contribute to a wetland project by a factor of three or more.

Asking a number of organizations to work alongside you can quickly build matching dollars.

Package your proposal to best meet the goals of the granting organization. One of the goals of the Wal-Mart Foundation is to improve air and water quality. Therefore, the foundation has chosen to fund wetland projects that are also designed to clean runoff before it reaches the sources of drinking water. A foundation interested in helping disadvantaged youth may look favorably at a wetland proposal that involves teens in a volunteer effort to remove noxious weeds and plant native aquatics as an excellent way to further its mission in your community.

Capitalize on the many benefits wetlands bring to your community. Funds may be available for cleaning water or for increasing tourism in an area, and wetland restoration can help achieve these goals.

Work alongside people, organizations, and businesses you know in your community. Businesses that begin by providing a small grant will often increase their contributions once they witness success.

Locating Sources for Funding

A number of free online services will alert you to the availability of grants from both government agencies and nonprofit organizations. You can fine-tune these services to provide you with only the alerts concerning environmental opportunities. The following three services are worth signing up for:

Federal government grant opportunities (www
 .grants.gov)
The Foundation Center http.//foundationcenter
 .org)
Eastern Coal Regional Roundtable (www
 .easterncoal.org)

Resources for Global Sustainability, Inc., annually publishes *Environmental Grantmaking Foundations* (http://www.environmentalgrants.com; 800-724-1857). The book contains information on over 900 foundations that fund environmental projects. I have found this publication most helpful in identifying organizations that may be interested in funding wetland projects, even though they do not have a history of such action.

Magazines that specialize in wetlands often feature stories about wetland restoration projects. These stories frequently list agencies, conservation organizations, and foundations that have provided funding for the wetland projects described. Scanning these stories may let you know who may be interested in funding your wetland project. Three important ones are the following:

Ducks Unlimited: Magazine available with a paid
 Ducks Unlimited membership; published
 by Ducks Unlimited, Inc., National Headquarters, One Waterfowl Way, Memphis, TN
 38120 (http://www.ducks.org)
Wetland Breaking News: Free newsletter published by the Association of State Wetland
 Managers (http://www.aswm.org)
The Nature Conservancy: Magazine and state
 newsletter available with a paid Nature Conservancy membership; published by the
 Nature Conservancey, Attn: Treasury (web
 support), 4245 North Fairfax Drive, suite 100,
 Arlington, VA 22203 (www.nature.org)

Federal Agencies

Private landowners interested in completing a wetland restoration project should contact a wildlife biologist who works for their respective state fish and wildlife agency to help identify possible government sources for funding. Many states offer programs where a biologist prepares a free management plan for privately owned lands. The plan opens the door for financial assistance from both federal and state agencies, and often the conservation work completed under such a plan is deductible from income for both federal and state income taxes.

A number of federal agencies are actively involved in financing wetland projects. Each offers a wide variety of programs that often include funding for public and private lands. Some of the most important are the Environmental Protection Agency (EPA), the Fish and Wildlife Service of the U.S. Department of the Interior (USDI), the National Oceanic and Atmospheric Administration (NOAA), and the Natural Resources Conservation Service (Wetland Reserve Program, Wildlife Habitat Incentives Program) of the U.S. Department of Agriculture (USDA).

State Agencies

Many states also fund wetland projects. Assistance can be available from a number of agencies such as fish and wildlife departments, divisions of water, divisions of forestry, and transportation departments. From time to time, states also receive grants from federal agencies that are designated for wetland projects.

Conservation Organizations

A number of conservation organizations finance wetland projects. The primary ones are the following:

Ducks Unlimited, Inc. (http://www.ducks.org): Ducks Unlimited raised over $180 million for wetlands and waterfowl conservation in fiscal year 2004, conserving nearly 210,000 acres of critical waterfowl habitat.[2] Working in partnership with federal, state, and county agencies, Ducks Unlimited will finance, design, and supervise the construction of wetlands with its international staff.

Delta Waterfowl (http://www.deltawaterfowl.org)

Waterfowl USA (http://www.waterfowlusa.org)

National Wildlife Federation (http://www.nwf.org): Offers Species Recovery Grants targeting habitat improvement for endangered species that may include wetland restoration and establishment

Wildlife Forever (http://www.wildlifeforever.org)

Foundations

Nearly 65,000 foundations in the United States gave away $29.7 billion in grants in 2003. The number of grant-giving foundations in the United States doubled from 1990 to 2002.[3] A majority of these organizations are required to give away 5 percent of their assets annually in order to maintain foundation status with the Internal Revenue Service. Each foundation has a mission that guides the projects it finances, and their missions can include helping the environment.

Foundations are very active in providing dollars for restoring wetlands throughout North America. The following nonprofit organizations are two that have provided funding for wetland restoration programs:

Wal-Mart Foundation (www.walmartfoundation.org): Emphasis on providing environmental grants to schools for environmental projects; proposals must be developed in cooperation with the Good Works coordinator at a local Wal-Mart, Sam's Club, Neighborhood Market, or Distribution Center Store, and then submitted by the store directly to the Wal-Mart Foundation

National Fish and Wildlife Foundation (http://www.nfwf.org): Dedicated to the conservation and management of fish, wildlife, plant resources, and the habitats on which they depend; the foundation administers a number of programs that include financing for wetland projects

Forming a Nonprofit Organization

Evelyn Morgan believes strongly in teaching people of all ages how they can help the environment. She has worked tirelessly as an interpreter for the Forest Service to organize workshops for educators, showing them how to use curriculum guides such as Project WET, Project Wild, Aquatic Project Wild, Project Learning Tree, and others. Not being happy with her annual budget, she wanted to find a legal and effective way to finance additional environmental education programs. To meet this need, she helped form the Sheltowee Environmental Education Coalition (SEEC) in 1998 in an effort to increase the effectiveness of the outdoor education program in the Daniel Boone National Forest. SEEC consisted of a small group of community volunteers who were dedicated to their cause.[4]

Evelyn talked me into sitting down with her to complete the forms required by the Internal Revenue Service (IRS) for SEEC to become a tax-exempt organization. Neither one of us had any experience in this area, so it took us a full day to get everything together for completing the forms. Fortunately, Evelyn had recently developed a constitution with bylaws for SEEC and had kept good financial records for the organization.

The IRS received our forms and sent us a letter with a few questions, showing understanding about our inexperience in this area. Within 30 days we received notification that SEEC was deemed exempt from federal income tax under section 501(a) of the Internal Revenue Code as an organization described in section 501(c)(3). I have since talked to a number of attorneys about the steps needed for an organization to receive 501(c)(3) status; they say that many attorneys offer their services free of charge to assist organizations to this end. So should you be afraid of doing the paperwork, perhaps you could muster the courage to ask an attorney for free help. Attorneys offering these pro bono services can also be a big help in incorporating and running your nonprofit organization.

Tony Burnett of the NRCS suggests that you ask an attorney and a CPA to serve on the board of your new 501(c)(3). Their volunteer advice and help at tax time can allow you to concentrate on helping the environment instead of filling out forms. Their skill can also be a big help as your organization grows and you have questions about contracting or the reimbursement of travel expenses.[5]

Becoming a tax-exempt organization provides important advantages when seeking funding for wetland restoration programs:

Most government agencies and nonprofit organizations will only grant funding to other tax-exempt organizations.

The donations given to tax-exempt organizations can generally be considered as tax deductions by the contributor.

The donations your organization receives are usually not taxed.

Designation can help lend legitimacy to your organization and its cause.

You can eliminate overhead or administrative costs by staffing your organization with volunteers, which, in turn, makes your cause more attractive to receive funding.

Obtaining tax-exempt status resulted in tremendous financial benefits for SEEC. The small organization has since received numerous grants from government agencies and other nonprofits to expand its environmental education efforts in the past 4 years. SEEC is now involved with wetland restoration, and has received five grants for this purpose in the past 3 years. Evelyn has now helped other groups receive tax-exempt status, including Project Learning Tree–Kentucky, Project Learning Tree–Ohio, Kentucky Envirothon, and even the Morehead Cattleman's Association.

Tony Burnett serves as coordinator of the Resource Conservation and Development Program (RC&D) for the Gateway RC&D Area, which encompasses nine counties in eastern Kentucky. He's the one who taught Evelyn and me how to obtain grants for environmental projects. RC&Ds are part of the NRCS, with the purpose of encouraging and improving the capability of local volunteer, elected, and civic leaders in designated RC&D areas to plan and carry out projects for resource conservation and community development. Their goals include fish and wildlife habitat projects and education projects, such as environmental education and outdoor classrooms, so wetland restoration and establishment fits right into their mission.[6]

Tony, like many other RC&D coordinators, is aware of government and nongovernment organizations that provide funding for wetland restoration. Another benefit of an RC&D is that it can serve as a 501(c)3 for projects adopted by the RC&D Council, being the conduit for transferring funds between the grant-giving organization and the receiving entity. With fourteen RC&D offices in Kentucky, and 375 throughout the United States as a whole, one may be located in your area.

Going after grant money can be very productive. Tony states: "It's possible for a good grant writer to bring in 100 times their salary a year." He says, "Once you come up with that initial contribution there's a snowball effect where the money will keep growing with each additional partner involved." Wetland projects that start with only a $500 contribution from a local wildlife club can grow with each additional organization that becomes involved until over $250,000 is available for a project.

Notes

Introduction

1. Henry F. French, *Farm Drainage* (New York: Orange Judd, 1903), 51.

2. L. M. Cowardin, V. Carter, F. C. Golet, and E. T. LaRoe, *Classification of Wetlands and Deepwater Habitats of the United States* (Washington, D.C.: U.S. Department of the Interior, Office of Biological Services, Fish and Wildlife Service, 1979).

1. Ages of Drainage

1. Quoted in Manly Miles, *Land Draining: A Handbook for Farmers on the Principles and Practice of Farm Draining* (New York: Orange Judd, 1893), 97–98.

2. Quoted in John H. Klippart, *The Principles and Practice of Land Drainage* (Cincinnati: Robert Clarke, 1861), 4.

3. Miles, *Land Draining,* 99.

4. Quoted in Marion M. Weaver, *History of Tile Drainage in America Prior to 1900* (Deposit, N.Y.: Valley Offset, 1964), 3.

5. John Johnstone, *An Account of the Mode of Draining Land,* 3rd edition (London: Richard Phillips, 1808), 134–135.

6. Peter C. Stewart, "The Shingle and Lumber Industries in the Great Dismal," *Journal of Forest History* (April 1981): 98–107.

7. Ann Vileisis, *Discovering the Unknown Landscape: A History of America's Wetlands* (Washington, D.C.: Island Press, 1997), 42.

8. Weaver, *History of Tile Drainage in America,* 9.

9. Agnes Hutchins Johnston, "Reminiscences of John Johnston, My Grandfather, of Geneva, New York, Originator of Tile Drainage in America." Contained in the printed program for the dedication of memorial to John Johnston, originator of tile drainage in America in 1935, by the American Society of Agricultural Engineers at Geneva, New York, October 9, 1935.

10. Weaver, *History of Tile Drainage in America,* 299.

11. "Test for the Expediency of Drainage," *Moore's Rural New Yorker* (Rochester, N.Y.), April 14, 1855.

12. Weaver, *History of Tile Drainage in America,* 221.

13. Ibid., 264.

14. Johnstone, *Account of the Mode of Draining Land,* 15.

15. Henry F. French, *Farm Drainage* (New York: Orange Judd, 1903), 310.

16. Charles Capen McLaughlin, *The Papers of Frederick Law Olmsted,* Volume 1, *The Formative Years 1822–1852* (Baltimore: Johns Hopkins University Press, 1977), 121, 339, 343; Sara Cedar Miller, *Central Park: An American Masterpiece* (New York: Harry N. Abrams, 2003), 75.

17. Charles G. Elliott, *Engineering for Land Drainage: A Manual for the Reclamation of Lands Injured by Water,* 3rd edition (London: Wiley, 1919), 16.

18. George E. Waring Jr., *Draining for Profit, and Draining for Health* (New York: Orange Judd, 1879), 88.

19. Ibid., 77.

20. French, *Farm Drainage,* 47.

21. Miller, *Central Park,* 13, 242; "Statement of the Quantity of Certain Classes of Work Done and of Materials Used in the Construction of Central Park, Exclusive of Operations on the Central Water Works of the City," Third Annual Report (New York: Department of Public Parks, 1874), 350–351.

22. Barry Lewis in "A Walk through Central Park," videotape (New York: Television Station 13/WNET, 2001).

23. Gary Dearborn, telephone conversations with author, New York, N.Y., March 1 and March 10, 2005.

24. Neil Calvanese, telephone conversation with author, New York, N.Y., March 14, 2005.

25. Quoted in Weaver, *History of Tile Drainage in America,* 48–49.

26. Crabs are believed to be the same as burrowing crayfish.

27. Quoted in *Thirteenth Annual Report of the American Institute of the City of New York for the year 1869–70* (Albany, N.Y.: Argus Printers, 1870), 785.

28. Klippart, *Principles and Practice of Land Drainage,* 177.

29. Quoted in Weaver, *History of Tile Drainage in America,* 142.

30. A *cat-swamp* was defined by clay tile maker D. Kenfield as a small pond containing water the greater part of the year and found usually on white oak ridges (Klippart, *Principles and Practice of Land Drainage,* 331–332). Kenfield also said that a cat-swamp had bluish-colored clay soils from 2 to 10 feet deep.

31. W. I. Chamberlain, *Tile Drainage; or Why, Where, When, and How to Drain Land with Tiles* (Medina, Ohio: A. I. Root, 1891), 124.

32. Quoted in French, *Farm Drainage,* 17–18.

33. Elliott, *Engineering for Land Drainage,* 244.

34. Weaver, *History of Tile Drainage in America,* 211.

35. Ibid., 222.

36. S. H. McCrory, "Historical Notes on Land Drainage in the United States," *Proceedings of the American Society of Civil Engineers* (1927), 53, as cited in Weaver, *History of Tile Drainage in America,* 227.

37. French, *Farm Drainage,* 361.

38. R. O. E. Davis, *Sponge Spicules in Swamp Soils,* Circular No. 67 (Washington, D.C.: USDA Bureau of Soils, 1912), 1; John R. Haswell, "Drainage in Humid Regions," in *Soils and Men: USDA Yearbook of Agriculture* (Washington, D.C.: U.S. Government Printing Office, 1938), 727.

39. Davis, *Sponge Spicules in Swamp Soils,* 1–2.

40. Haswell, "Drainage in Humid Regions," 727.

41. Davis, *Sponge Spicules in Swamp Soils,* 4.

42. Haswell, "Drainage in Humid Regions," 727.

43. John Keenon, telephone conversation with author, Clay City, Ky., January 18, 2004.

44. Don Hurst, field interview by author, Stanton, Ky., December 2, 2003.

45. Vileisis, *Discovering the Unknown Landscape,* 301.

46. Randy Smallwood, field interview by author, Owingsville, Ky., June 27, 2003.

47. U.S. Department of Agriculture, *Wildlife Habitat Incentives Program Fact Sheet: Farm Bill 2002* (Washington, D.C.: U.S. Government Printing Office, 2004), 1–2.

2. Why They Pulled the Plug

1. For examples, see Manly Miles, *Land Draining: A Handbook for Farmers on the Principles and Practice of Farm Draining* (New York: Orange Judd, 1893); 2, Henry F. French, *Farm Drainage* (New York.: Orange Judd, 1903), 97; John Johnstone, *An Account of the Mode of Draining Land,* 3rd edition (London: Richard Phillips, 1808), 3, 115; Marion M. Weaver, *History of Tile Drainage in America Prior to 1900* (Deposit, N.Y.: Valley Offset, 1964), 264.

2. Sir Charles Bart Coote, *General View of the Agriculture and Manufactures of the Kings County, with Observations on the Means of Improvement: Drawn up in the Year 1801 for the Consideration, and Under the Direction of the Dublin Society* (Dublin: Graisberry and Campbell, 1801), 129.

3. Quoted in Weaver, *History of Tile Drainage in America,* 240.

4. Wayne Pettit, telephone conversation with author, Morehead, Ky., March 8, 2005.

5. John R. Haswell, "Drainage in Humid Regions," in *Soils and Men: USDA Yearbook of Agriculture* (Washington, D.C.: U.S. Government Printing Office, 1938), 723.

6. W. I. Chamberlain, *Tile Drainage; or Why, Where, When, and How to Drain Land with Tiles* (Medina, Ohio: A. I. Root, 1891), 13; John H. Klippart, *The Principles and Practice of Land Drainage* (Cincinnati: Robert Clarke, 1861), 173–176; French, *Farm Drainage,* 91.

7. Klippart, *Principles and Practice of Land Drainage,* 128–129; Weaver, *History of Tile Drainage in America,* 32, 93, 263.

8. Charles G. Elliott, *Engineering for Land Drainage: A Manual for the Reclamation of Lands Injured by Water,* 3rd edition (London: Wiley, 1919), 63; George E. Waring Jr., *Draining for Profit, and Draining for Health* (New York: Orange Judd, 1879), 31; Klippart, *Principles and Practice of Land Drainage,* 72, 117; French, *Farm Drainage,* 89; Miles, *Land Draining,* 74; Weaver, *History of Tile Drainage in America,* 249.

9. Chamberlain, *Tile Drainage,* 13; French, *Farm Drainage,* 92; Elliott, *Engineering for Land Drainage,* 63, Weaver, *History of Tile Drainage in America,* 28.

10. Klippart, *Principles and Practice of Land Drainage,* 89, 129; Waring, *Draining for Profit, and Draining for Health,* 34; Elliott, *Engineering for Land Drainage,* 63; Miles, *Land Draining,* 26.

11. Chamberlain, *Tile Drainage,* 8; Waring, *Draining for Profit, and Drainage for Health,* 66–67; Klippart, *Principles and Practice of Land Drainage,* 98; French, *Farm Drainage,* 258; Miles, *Land Draining,* 8.

12. Chamberlain, *Tile Drainage,* 19; Klippart, *Principles and Practice of Land Drainage,* 89, 112–118; French, *Farm Drainage,* 67; Miles, *Land Draining,* 74.

13. Klippart, *Principles and Practice of Land Drainage,* 112–118; Weaver, *History of Tile Drainage in America,* 28, 234.

14. Chamberlain, *Tile Drainage,* 15, 25; Klippart, *Principles and Practice of Land Drainage,* 89–90, 147–152; Waring, *Draining for Profit, and Draining for Health,* 168; French, *Farm Drainage,* 75–77; Elliott, *Engineering for Land Drainage,* 63–64; Miles, *Land Draining,* 25, 74.

15. Chamberlain, *Tile Drainage,* 15–16; Klippart, *Principles and Practice of Land Drainage,* 134–143, 165–170; French, *Farm Drainage,* 103.

16. Klippart, *Principles and Practice of Land Drainage,* 175–176; French, *Farm Drainage,* 95–96; Elliott, *Engineering for Land Drainage,* 64; Weaver, *History of Tile Drainage,* 31.

17. Chamberlain, *Tile Drainage,* 27; Waring, *Draining for Profit, and Draining for Health,* 208–215; French, *Farm Drainage,* 16; Weaver, *History of Tile Drainage in America,* 242.

18. Chamberlain, *Tile Drainage,* 18, 22; Klippart, *Principles and Practice of Land Drainage,* 90; Elliott, *Engineering for Land Drainage,* 63–64; French, *Farm Drainage,* 67; Miles, *Land Draining,* 74; Weaver, *History of Tile Drainage in America,* 93.

19. Chamberlain, *Tile Drainage,* 26; Klippart, *Principles and Practice of Land Drainage,* 172; Miles, *Land Draining,* 74; Weaver, *History of Tile Drainage in America,* 255.

20. Chamberlain, *Tile Drainage,* 41–43; Weaver, *History of Tile Drainage in America,* 248.

21. George McClure, telephone conversations with author, January 2 and 4, 2005; also, field interview by author, Wheelersburg, Ohio, January 8, 2005.

22. Klippart, *Principles and Practice of Land Drainage,* 179–180.

23. Charles Gleason Elliott, *Practical Farm Drainage* (New York: Wiley, 1903), 3.

24. Miles, *Land Draining,* 25.

25. Ibid., 70–71; Weaver, *History of Tile Drainage in America,* 67.

26. Klippart, *Principles and Practice of Land Drainage,* 181.

27. French, *Farm Drainage,* 93.

28. T. W. Edminister and Ronald C. Reeve, "Drainage Problems and Methods," in *Soil: USDA Yearbook of Agriculture* (Washington, D.C.: U.S. Government Printing Office, 1957), 382.

29. E. A. Schlaudt, "Drainage in Forestry Management in the South," in *Water: USDA Yearbook of Agriculture* (Washington, D.C.: U.S. Government Printing Office, 1955), 564.

30. Ibid., 565.

31. Earl L. Stone Jr. and Paul E. Lemmon, "Soil and the Growth of Forests," in *Soil: USDA Yearbook of Agriculture* (Washington, D.C.: U.S. Government Printing Office, 1957), 725.

32. Frank W. Woods, Otis L. Copeland Jr., and Carl E. Ostrom, "Soil Management for Forest Trees," in *Soil: USDA Yearbook of Agriculture* (Washington, D.C.: U.S. Government Printing Office, 1957), 712.

33. Schlaudt, "Drainage in Forestry Management in the South," 565.

34. Ibid.

35. Ibid., 567.

36. Woods, Copeland, and Ostrom, "Soil Management for Forest Trees," 712.

37. Schlaudt, "Drainage in Forestry Management in the South," 567.

38. James Duininck, vice president of sales, Prinsco Inc., telephone conversation with author, Prinsburg, Minn., August 22, 2005; Charlie Schafer, president, Agri-Drain Corp., interview by author, Kansas City, Mo., March 9. 2004.

3. Ditching for Dollars

1. Henry F. French, *Farm Drainage* (New York: Orange Judd, 1903), 80–81, 99–100.

2. John H. Klippart, *The Principles and Practice of Land Drainage* (Cincinnati: Robert Clarke, 1861), 100–102; French, *Farm Drainage,* 361–363.

3. Charles G. Elliott, *Engineering for Land Drainage: A Manual for the Reclamation of Lands Injured by Water,* 3rd edition (London: Wiley, 1919), 162.

4. Quincy A. Ayres and Daniel Scoates, *Land Drainage and Reclamation* (New York: McGraw-Hill, 1928), 107, 109.

5. Ibid., 130–131.

6. George McClure, field interview by author, Wheelersburg, Ohio, January 8, 2005.

7. French, *Farm Drainage,* 99.

8. David Murphy, field interview by author, Farmers, Ky., March 8, 2005.

9. Robert W. Burwell and Lawson G. Sugden, "Potholes-Going, Going . . . ," in *Waterfowl Tomorrow,* edited by Joeph Linduska (Washington, D.C.: U.S. Department of the Interior, U.S. Fish and Wildlife Service, 1964), 370.

10. Quoted in Marion M. Weaver, *History of Tile Drainage in America Prior to 1900* (Deposit, N.Y.: Valley Offset, 1964), 224.

11. W. I. Chamberlain, *Tile Drainage; or Why, Where, When, and How to Drain Land with Tiles* (Medina, Ohio: A. I. Root, 1891), 61.

12. Don Hurst, field interview by author, Stanton, Ky., December 2, 2003.

13. French, *Farm Drainage,* 101–103.

14. "Draining Low Land," *Moore's Rural New Yorker* (Rochester, N.Y.), May 5, 1855.

15. Klippart, *Principles and Practice of Land Drainage,* 177.

16. Weaver, *History of Tile Drainage in America,* 162.

17. Elliott, *Engineering for Land Drainage,* 221.

18. Ricky Wells and Randy Smallwood, field interview by author, Owingsville, Ky., June 27, 2003.

19. Don Hurst, telephone conversation with author, Stanton, Ky., November 20, 2003.

20. Ayres and Scoates, *Land Drainage and Reclamation,* 202, 225–226, 230.

21. Richard Bond, telephone conversations with author, December 7, 2004, and March 23, 2005; field interview by author, Grayson, Ky., December 9, 2004.

22. John Johnstone, *An Account of the Mode of Draining Land,* 3rd edition (London: Richard Phillips, 1808), 182–183.

23. French, *Farm Drainage,* 99.

24. McClure, field interview; also, telephone conversations with author, January 2 and 4, 2005.

25. Bond, field interview.

26. Allan T. Studholme and Thomas Sterling, "Dredges and Ditches," in *Waterfowl Tomorrow,* edited by Joseph Linduska (Washington, D.C.: U.S. Department of the Interior, U.S. Fish and Wildlife Service, 1964), 360.

4. Plowing to Drain

1. John Newman, field interview by author, Wallingford, Ky., January 5, 2005; also, telephone conversations, January 6, 2005, August 18, 2006, and September 11, 2006.

2. Manly Miles, *Land Draining: A Handbook for Farmers on the Principles and Practice of Farm Draining* (New York: Orange Judd, 1893), 100–101.

3. John Johnstone, *An Account of the Mode of Draining Land,* 3rd edition (London: Richard Phillips, 1808), 138, 160–161, 170–172.

4. Henry F. French, *Farm Drainage* (New York: Orange Judd, 1903), 101.

5. W. I. Chamberlain, *Tile Drainage; or Why, Where, When, and How to Drain Land with Tiles* (Medina, Ohio: A. I. Root, 1891), 56.

6. John H. Klippart, *The Principles and Practice of Land Drainage* (Cincinnati: Robert Clarke, 1861), 182.

7. Quincy A. Ayres and Daniel Scoates, *Land Drainage and Reclamation* (New York: McGraw-Hill, 1928), 321.

8. John R. Haswell, "Drainage in Humid Regions," in *Soils and Men: USDA Yearbook of Agriculture* (Washington, D.C.: U.S. Government Printing Office, 1938), 726.

9. George McClure, field interview by author, Wheelersburg, Ohio, January 8, 2005; also, telephone conversations with author, January 2 and 4, 2005.

10. Haswell, "Drainage in Humid Regions," 734.

11. Klippart, *Principles and Practice of Land Drainage,* 231–232; French, *Farm Drainage,* 159.

12. Klippart, *Principles and Practice of Land Drainage,* 245–247.

13. French, *Farm Drainage,* 109.

14. Klippart, *Principles and Practice of Land Drainage,* 247.

15. Haswell, "Drainage in Humid Regions," 733–734.

16. John Utterback, field interview by author, Cranston, Ky., August 2, 2004.

5. Sticks and Stones

1. John Johnstone, *An Account of the Mode of Draining Land,* 3rd edition (London: Richard Phillips, 1808), 65, 90; Henry F. French, *Farm Drainage* (New York: Orange Judd, 1903), 111–112.

2. Marion M. Weaver, *History of Tile Drainage in America Prior to 1900* (Deposit, N.Y.: Valley Offset, 1964), 49, 267.

3. John Newman, field interview by author, Wallingford, Ky., January 5, 2005; also, telephone conversations, January 6, 2005, August 18, 2006, and September 11, 2006.

4. Jack D. Ellis, *Morehead Memories: True Stories from Eastern Kentucky* (Ashand, Ky.: Jesse Stuart Foundation, 2001), 268.

5. John Meredith, field interview by author, Jackson, Ohio, January 8, 2005.

6. Ed Stevens, field interview by author, Grayson, Ky., December 7, 2004.

7. French, *Farm Drainage,* 104–105; John H. Klippart, *The Principles and Practice of Land Drainage* (Cincinnati: Robert Clarke, 1861), 222–226; Weaver, *History of Tile Drainage in America,* 49–50.

8. French, *Farm Drainage,* 112.

9. Richard Bond, field interview by author, Grayson, Ky., December 9, 2004; also, telephone conversations with author, December 7, 2004, and March 23, 2005.

10. Richard Neal, field interview by author, Lecta, Ohio, July 27, 2004.

11. Meredith, field interview.

12. George McClure, field interview by author, Wheelersburg, Ohio, January 8, 2005; also, telephone conversations with author, January 2 and 4, 2005.

13. Johnstone, *Account of the Mode of Draining Land,* 155–158.

14. French, *Farm Drainage,* 112.

15. Bond, field interview and telephone conversations.

16. Sir Charles Bart Coote, *General View of the Agriculture and Manufactures of the Kings County, with Observations on the Means of Improvement: Drawn up in the Year 1801 for the Consideration, and Under the Direction of the Dublin Society* (Dublin: Graisberry and Campbell, 1801), 71.

17. Henry Stephens, *Book of the Farm,* Volume 1 (New York: Greeley and McElrath, 1847), 318–320.

18. Klippart, *Principles and Practice of Land Drainage,* 17–20.

19. Johnstone, *Account of the Mode of Draining Land,* 122.

20. Ibid., A2–A4.

21. Ibid., 19.

22. French, *Farm Drainage,* 365; Charles G. Elliott, *Engineering for Land Drainage: A Manual for the Reclamation of Lands Injured by Water,* 3rd edition (London: Wiley, 1919), 77–78, 310; Klippart, *Principles and Practice of Land Drainage,* 24–25.

23. French, *Farm Drainage,* 34–35.

24. I. Whiting, "Underdraining: Stone and Tile," *Moore's Rural New Yorker* (Rochester, New York), March 31, 1855, 102.

25. Quoted in Weaver, *History of Tile Drainage in America,* 265–266.

26. Klippart, *Principles and Practice of Land Drainage,* 258.

27. Ibid., 259–260.

28. French, *Farm Drainage,* 223.

29. Bond, field interview and telephone conversations.

30. Edward Ratcliff, field interview by author, Grayson, Ky., December 30, 2004.

31. French, *Farm Drainage,* 117.

32. Scott Manning, telephone conversation with author, Flemingsburg, Ky., August, 2, 2004.

33. Meredith, field interview.

34. McClure, field interview and telephone conversations.

6. Miles of Tiles

1. W. I. Chamberlain, *Tile Drainage; or Why, Where, When, and How to Drain Land with Tiles* (Medina, Ohio: A. I. Root, 1891), 3.

2. Manly Miles, *Land Draining: A Handbook for Farmers on the Principles and Practice of Farm Draining* (New York: Orange Judd, 1893), 139.

3. Henry F. French, *Farm Drainage* (New York: Orange Judd, 1903), xi.

4. Chamberlain, *Tile Drainage,* 4.

5. Ibid., 13.

6. Marion M. Weaver, *History of Tile Drainage in America Prior to 1900* (Deposit, N.Y.: Valley Offset, 1964), 245; John H. Klippart, *The Principles and Practice of Land Drainage* (Cincinnati: Robert Clarke, 1861); 3.

7. Quoted in Weaver, *History of Tile Drainage in America,* 144–145.

8. Chamberlain, *Tile Drainage,* 59.

9. Miles, *Land Draining,* 183–184.

10. Weaver, *History of Tile Drainage in America,* 132.

11. George McClure, field interview by author, Wheelersburg, Ohio, January 8, 2005; also, telephone conversations with author, January 2 and 4, 2005.

12. Quoted in Weaver, *History of Tile Drainage in America,* 199–200.

13. French, *Farm Drainage,* 175.

14. James Flowers, field interview by author, Morgantown, Ky., November 11, 2004.

15. Philip Annis, telephone conversation with author, Lexington, Ky., May 8, 2004; also, field interview, Morgantown, Ky., November 11, 2004.

16. Weaver, *History of Tile Drainage in America,* 321.

17. Keith H. Beauchamp, "Tile Drainage: Its Installation and Upkeep," in *Water: USDA Yearbook of Agriculture* (Washington, D.C.: U.S. Government Printing Office, 1955), 508–509.

18. Klippart, *Principles and Practice of Land Drainage,* 296, 299.

19. French, *Farm Drainage,* 169.

20. George E. Waring Jr., *Draining for Profit, and Draining for Health* (New York: Orange Judd, 1879), 74.

21. Weaver, *History of Tile Drainage in America,* 187–188.

22. Richard Bond Jr., field interview by author, Grayson, Ky., December 9, 2004; also, telephone conversations with author, December 7, 2004, and March 23, 2005.

23. Chamberlain, *Tile Drainage,* 63.

24. Klippart, *Principles and Practice of Land Drainage,* 381.

25. Ibid., 375–378.

26. McClure, field interview and telephone conversations.

27. John R. Haswell, "Drainage in Humid Regions," in *Soils and Men: USDA Yearbook of Agriculture* (Washington, D.C.: U.S. Government Printing Office, 1938), 731.

7. Massive Machines and Plastic Drain Lines

1. Ed Stevens and Jimmy Lyons, field interview by author, Grayson, Ky., December 7, 2004.

2. Donna M. Back, "Tile Machine Works in Rowan District," *Soil Conservation District Activities Newsletter* (Morehead, Ky.), 4th Quarter, 1977.

3. Richard Bond Jr., field interview by author, Grayson, Ky., December 9, 2004; also, telephone conversations with author, December 7, 2004, and March 23, 2005.

4. George McClure, field interview by author, Wheelersburg, Ohio, January 8, 2005; also, telephone conversations with author, January 2 and 4, 2005.

5. George E. Waring Jr., *Draining for Profit, and Draining for Health* (New York: Orange Judd, 1879), 98–100.

6. Quoted in Weaver, *History of Tile Drainage in America,* 202–203.

7. Charles G. Elliott, *Engineering for Land Drainage: A Manual for the Reclamation of Lands Injured by Water,* 3rd edition (London: Wiley 1919), 151–153.

8. Jimmy Lyons, telephone conversation with author, Grayson, Ky., January 5, 2004; also, field interview by author, December 3, 2004, and office interview by author, December 9, 2004.

9. P. Comer, K. Goodin, G. Hammerson, S. Menard, M. Pyne, M. Reid, M. Robles, M. Russo, L. Sneddon, K. Snow, A. Tomaino, and M. Tuffly, *Biodiversity Values of Geographically Isolated Wetlands: An Analysis of 20 U.S. States* (Arlington, Va.: NatureServe, 2005), 7.

10. Manly Miles, *Land Draining: A Handbook for Farmers on the Principles and Practice of Farm Draining* (New York: Orange Judd, 1893), 186; Quincy A. Ayres and Daniel Scoates, *Land Drainage and Reclamation* (New York: McGraw-Hill, 1928), 393.

11. John Johnstone, *An Account of the Mode of Draining Land,* 3rd edition (London: Richard Phillips, 1808), 63–65.

12. John H. Klippart, *The Principles and Practice of Land Drainage* (Cincinnati: Robert Clarke, 1861), 102–103.

13. Ayres and Scoates, *Land Drainage and Reclamation,* 393–394.

14. Johnstone, *Account of the Mode of Draining Land,* 117.

15. Henry F. French, *Farm Drainage* (New York.: Orange Judd, i903), 85.

16. Ibid., 366.

17. Ibid., 85–87, 341, 348–349.

8. Filling and Leveling

1. John Johnstone, *An Account of the Mode of Draining Land,* 3rd edition (London: Richard Phillips, 1808), 191–192.

2. W. I. Chamberlain, *Tile Drainage; or Why, Where,*

When, and How to Drain Land with Tiles (Medina, Ohio: A. I. Root, 1891), 129.

3. Robert W. Burwell and Lawson G. Sugden, "Potholes—Going, Going . . . ," in *Waterfowl Tomorrow,* edited by Joseph Linduska (Washington, D.C.: U.S, Department of the Interior, U.S. Fish and Wildlife Service, 1964), 371.

4. Jimmy Lyons, telephone conversation with author, Grayson, Ky., January 5, 2004; also, field interview by author, December 3, 2004, and office interview by author, December 9, 2004.

5. Richard Bond Jr., field interview by author, Grayson, Ky., December 9, 2004; also, telephone conversations with author, December 7, 2004, and March 23, 2005.

6. Earl J. Osborne, field interview by author, West Liberty, Ky., February 17, 2004.

7. Ibid.

8. Loren M. Smith, *Playas of the Great Plains* (Austin: University of Texas Press, 2003), 166.

9. Anonymous, telephone conversations with and field interviews by author from August 1, 1988, to October 1, 2007.

9. Wetland Drainage Stories

1. Thomas E. Dahl, *Wetlands Losses in the United States, 1780's to 1980's* (Washington, D.C.: U.S. Department of the Interior, Fish and Wildlife Service, 1990), 6.

2. Randy Smallwood, telephone conversation with author, Owingsville, Ky., June 20, 2003; also, office interview by author, September 12, 2003, and field interview by author, November 10, 2003.

3. Rand and Cindy Ragland, field interview by author, Ellards, Ala., September 18–19, 2003; also, e-mail correspondence, December 15, 2003.

4. John D. Smith, field interviews by author, Sudith, Ky., September 29, 2003, October 6, 2003, August 25, 2005, and September 12, 2006.

5. Wes Tuttle, field interview by author, Sudith, Ky., August 25, 2005.

6. Smith. field interviews.

7. Merlin and Greg Spencer, field interviews by author, Frenchburg, Ky., April 25, 2005, June 30, 2005, August 25, 2005, September 12, 2006; also, telephone conversations with author, April 4, 11, and 27, 2005, and September 11, 2006.

8. R. B. Gray, "Tillage Machinery," in *Soils and Men: USDA Yearbook of Agriculture* (Washington, D.C.: U.S. Government Printing Office, 1938), 331.

9. Kim Spencer, field interview by author, Frenchburg, Ky., April 25, 2005.

10. Wes Tuttle, field interview by author, Frenchburg, Ky., August 25, 2005.

11. Agnes Hutchins Johnston, "Reminiscences of John Johnston, My Grandfather, of Geneva, New York,

Originator of Tile Drainage in America." Contained in the printed program for the dedication of memorial to John Johnston, originator of tile drainage in America in 1935, by the American Society of Agricultural Engineers at Geneva, New York, October 9, 1935, 5.

12. John R. Haswell, "Drainage in Humid Regions," in *Soils and Men: USDA Yearbook of Agriculture* (Washington, D.C.: U.S. Government Printing Office, 1938), 729.

13. Marion M. Weaver, *History of Tile Drainage in America Prior to 1900* (Deposit, N.Y.: Valley Offset, 1964), 18.

14. Merrill Roenke, field interview by author, Geneva, N.Y., August 21, 2003; also, telephone conversation with author, December 4, 2003. Eddy Kime, telephone conversation with author, Geneva, N.Y., December 5, 2003.

15. Larry Cecere, telephone conversation with author, Geneva, N.Y., December 4, 2003. Roenke, telephone conversation.

16. Kime, telephone conservation.

17. *Soil Survey of Seneca County New York* (Washington, D.C.: Soil Conservation Service, U.S. Department of Agriculture, 1972), 121.

18. Melissa Yearick, e-mail correspondence with author, Horsehead, N.Y., February 24, 2005.

19. Earl J. Osborne, field interview by author, West Liberty, Ky., February 17, 2004.

20. David Taylor, field interview by author, Corbin, Ky., March 22, 2004; also, telephone conversation, March 23, 2004.

21. Donnie Centers and Billy Osborne, field interviews by author, Hazel Greene, Ky., July 8, 2004.

22. Jimmy Lyon, field interview by author, Grayson, Ky., December 3, 2004.

23. Scott Whitecross, field interview by author, 100 Mile House, B.C., July 26, 2004; also, letter to author, July 26, 2004.

10. Vestiges of Wetlands Long Past

1. Don Hurst, field interview by author, Stanton, Ky., December 2, 2003; also, telephone conversation with author, November 20, 2003.

2. H. H. Wooten and L. A. Jones, "The History of Our Drainage Enterprises," *Yearbook of Agriculture* 32 (1955): 478–491.

3. John H. Klippart, *The Principles and Practice of Land Drainage* (Cincinnati: Robert Clarke, 1861), 319.

4. Richard Bond Jr., field interview by author, Grayson, Ky., December 9, 2004; also. telephone conversations with author, December 7, 2004, and March 23, 2005.

5. Henry F. French, *Farm Drainage* (New York: Orange Judd, 1903), 94.

6. Ibid., 100.

7. Doreen Miller, field interview by author, Franklin, N.C., May 20, 2004, and October 27–28, 2004; also, e-mail correspondence with author, December 20, 2004.

8. Charles G. Elliott, *Engineering for Land Drainage: A Manual for the Reclamation of Lands Injured by Water,* 3rd edition (London: Wiley, 1919), 322–328.

9. John Johnstone, *An Account of the Mode of Draining Land,* 3rd edition (London: Richard Phillips, 1808), 86–87.

10. W. I. Chamberlain, *Tile Drainage; or Why, Where, When, and How to Drain Land with Tiles* (Medina, Ohio: A. I. Root, 1891), 126–127.

11. Elliott, *Engineering for Land Drainage,* 237–238.

12. French, *Farm Drainage,* 372.

13. Chamberlain, *Tile Drainage,* 131–132.

14. James "Booster" Flowers, field interview by author, Morgantown, Ky., November 11, 2004.

15. Manly Miles, *Land Draining: A Handbook for Farmers on the Principles and Practice of Farm Draining* (New York: Orange Judd, 1893), 184.

16. French, *Farm Drainage,* 316.

17. Alice L. Thompson and Charles S. Luthin, *Wetland Restoration Handbook for Wisconsin Landowners,* 2nd edition (Madison: Bureau of Integrated Science Services, Wisconsin Department of Natural Resources, 2004), 41.

18. Klippart, *Principles and Practice of Land Drainage,* 249–250; Elliott, *Engineering for Land Drainage,* 326–327.

19. French, *Farm Drainage,* 320–321.

20. *2002 Annual NRI,* Natural Resources Inventory, U.S. Department of Agriculture, 2004, at http://www.nrcs.usda.gov/technical/NRI/ (retrieved February 20, 2005).

21. "Hydric Soils of the United States," *Federal Register* 60, no. 37 (February 24, 1995): 10349.

22. J. L. Richardson and M. J. Vepraskas, *Wetland Soils* (Boca Raton, Fla.: CRC Press, 2001), 31.

23. Ibid., 19, 198.

24. Ibid., 192–193.

25. Environmental Laboratory, *Corps of Engineers Wetlands Delineation Manual,* Technical Report Y-87–1 (Vicksburg, Miss.: U.S. Army Engineer Waterways Experiment Station, 1987), 26.

26. J. W. Stucki, B. A. Goodman, and U. Schwertmann, *Iron in Soils and Clay Minerals* (Dordrecht: Reidel, 1985), 813–814.

27. Richardson and Vepraskas, *Wetland Soils,* 171.

28. Stucki, Goodman, and Schwertmann, *Iron in Soils and Clay Minerals,* 813–814.

29. Ibid., 811.

30. Richardson and Vepraskas, *Wetland Soils,* 173, 169.

31. Thomas E. Dahl, *Wetlands Losses in the United States, 1780's to 1980's* (Washington, D.C.: U.S. Department of the Interior, Fish and Wildlife Service, 1990), 8–9.

32. Hinrich L. Bohn, Brian L. McNeal, and George A. O'Connor, *Soil Chemistry,* 2nd edition (New York: Wiley, 1985), 272.

33. Philippine Ministry of Agriculture, *Wetland Soils: Characterization, Classification, and Utilization* (Los Baños, Laguna, Philippines: International Rice Research Institute, 1985), 62.

34. Bond, field interview and telephone conversations.

35. Marty McCleese, office interview by author, Morehead, Ky., March 22, 2005.

11. Building and Restoration

1. Susan-Marie Stedmen, *An Introduction and User's Guide to Wetland Restoration, Creation, and Enhancement* (Washington, D.C.: U.S. Environmental Protection Agency, 2003), at http://www.epa.gov/owow/wetlands/pdf/restdocfinal.pdf (accessed October 27, 2006), 29.

2. Don Hurst, field interview by author, Stanton, Ky., December 2, 2003; Mark Lindflott, field interview by author, Des Moines, Iowa, March 9, 2004, and telephone conversation with author, April 20, 2004.

3. Henry F. French, *Farm Drainage* (New York.: Orange Judd, 1903), 322–332.

4. R. J. Naiman, J. M. Melillo, and J. E. Hobbie, "Ecosystem Alteration of Boreal Forest Streams by Beaver (*Castor canadensis*)," *Ecology* 67 (1986): 1254–1269; R. J. Naiman, C. A. Johnston, and J. C. Kelly, "Alteration of North American Streams by Beaver," *BioScience* 38 (1988): 753–762.

5. F. J. Singer, D. Labrode, and L. Sprague, *Beaver Reoccupation and an Analysis of the Otter Niche in Great Smoky Mountains National Park,* Report No. 40 (Atlanta: National Park Service, Southeast Regional Office Research/Resource Management, 1981).

6. Doreen Miller, field interview by author, Franklin, N.C., May 20, 2004, and October 27–28, 2004; also, e-mail correspondence with author, December 20, 2004.

7. L. Wilsson, "Observations and Experiments on the Ethology of the European Beaver (*Castor fiber L.*)," *Viltrevy* 8 (1971): 115–266; A. M. Hartman, "Analysis of Conditions Leading to the Regulation of Water Flow by a Beaver," *Psychological Research* 25 (1975): 427–431.

8. James Curatolo, field interview by author, Horseheads, N.Y., August 22, 2003.

9. K. J. Babbit and G. W. Tanner, "Use of Temporary Wetlands by Anurans in a Hydrologically Modified Landscape," *Wetlands* 20, no. 2 (2000): 313–322.

10. Environmental Laboratory, *Corps of Engineers Wetlands Delineation Manual,* Technical Report Y-87–1 (Vicksburg, Miss.: U.S. Army Engineer Waterways Experiment Station, 1987).

11. Kathleen Kuna, telephone conversation with author, Nashville, Tenn., January 10, 2006.

12. Nationwide Permit No. 27, *Federal Register* 61, no. 241 (December 13, 1996), 65917; modified, *Federal Register* 67, no. 10 (January 15, 2002), 2082.

13. *Federal Register* 67, no 10 (January 15, 2002), 2082.

14. Lisa Morris, telephone conversation with author, Nashville, Tenn., January 10, 2006.

15. *Regulatory Program Mission Statement,* U.S. Army Corps of Engineers, July 12, 2006, at http://www.usace.army.mil/cw/cecno/reg_faq.htm waste (retrieved October 27, 2006), 2.

16. Craig LeSchack, telephone conversations and e-mail correspondence with author, Charleston, S.C., April 3 and 6, 2006, June 20, 2006, July 31, 2006, and February 5, 2007.

17. Danny Fraley, telephone conservation with author, Morehead, Ky., January 10, 2006.

18. "Standards and Guidelines for Archaeology and Historic Preservation," *Federal Register* 48, no. 190 (September 29, 1983).

19. Advisory Council on Historic Preservation, 36 CFR Part 800, Protection of Historic Properties.

20. Ibid., secs. 4[b][1], 4[c], and 5.

21. Frank Bodkin, interviews by author, Morehead, Ky., January 2006.

22. 36 CFR Part 800, sec.16[d].

23. Ibid., subpart B, sec. 9 [b].

24. Bodkin, interviews.

25. *Conserving America's Wetlands, Implementing the President's Goal* (Washington, D.C.: Council on Environmental Quality, Executive Office of the President, 2005).

26. Mike Armstrong, telephone conversation with author, Frankfort, Ky., January 10, 2006.

12. Building Wetlands on Dry Land

1. John Johnstone, *An Account of the Mode of Draining Land,* 3rd edition (London: Richard Phillips, 1808), 23–24.

2. Kathy Flegel, field interviews by author, as well as telephone conversations and e-mail correspondence with author, Ironton, Ohio, 1993–2002.

3. Kathy Flegel, field interviews by author, as well as telephone conversations and e-mail correspondence with author, Ironton, Ohio, 2003–2005.

4. Eddie Park, field interviews by and telephone conversations with author, Ironton, Ohio, 2004–2006.

5. Doug Malsam, telephone conversation with author, WaKeeney, Kan., May 3, 2004; Bud Malsam, telephone conversation with author, WaKeeney, Ks., May 5, 2004.

6. Brad Feaster, telephone conversations with author, Mt. Vernon, Ind., April 19 and 20, 2004; also, e-mail correspondence with author, April 20, 2004, and May 18, 2004.

7. Brad Feaster, letter to author, Mt. Vernon, Ind., May 18, 2004.

8. Tommy Counts, telephone conversations and e-mail correspondence with author, Double Springs, Ala., March 13, 2004, and May 10, 2004.

9. Donald Back, field interview by author, Frenchburg, Ky., 1993.

10. Richard Hunter, field interview by author, Frenchburg, Ky., 1993.

11. Ron Taylor, field interview by author, Morehead, Ky., April 14, 2005.

12. Dennis Eger, telephone conversations and e-mail correspondence with author, Montgomery, Ind., June 24 and 25, 2004, October 8 and 12, 2004, and December 1 and 10, 2004. Pat Merchant, telephone conversation and e-mail correspondence with author, Bedford, Ind., June 28, 2004.

13. Dennis Eger, telephone conversation with author, Montgomery, Ind., October 23, 2006.

14. A. J. K. Calhoun and P. deMaynadier, *Forestry Habitat Management Guidelines for Vernal Pool Wildlife,* Technical Paper No. 6 (Bronx, N.Y.: Metropolitan Conservation Alliance, Wildlife Conservation Society, 2004), 11.

15. Thomas R. Biebighauser, *A Guide to Creating Vernal Ponds* (Washington, D.C.: Forest Service, U.S. Department of Agriculture, 2003).

13. Building Wetlands on Wet Land

1. Barb Leuelling, telephone conversation with author, Duluth, Minn., August 26, 2004; also, e-mail correspondence, August 9 and 30, 2004, and December 20 and 22, 2004. Barb interviewed Stu Behling on my behalf.

2. Michael E. Kenawell, "Patterns of Avifauna Use of Constructed Wetlands in the Beaver Creek Wetland Complex, Menifee County, Kentucky," Master of Science thesis, Morehead State University, Morehead, Kentucky, 2002.

3. Paul Tine', telephone conversation with author, Brimson, Minn., March 2, 2004; also, e-mail correspondence with author, April 22, 2004, and June 8, 2004, and letter to author, July 26, 2004. Denny FitzPatrick, e-mail correspondence with author, Grand Marais, Minn., July 2, 7, 13, and 14, 2004. Jon Hakala, telephone conversations with author, Ely, Minn., April 22, 2002, June 21 and 28, 2004, and October 13, 2004; also, e-mail correspondence with author and letter to author, June 29, 2004.

4. Tine', e-mail correspondence and letter.

5. Hakala, telephone conversations.

6. Charles Korson and Christopher Pearl, seeing spots in a cascades meadow, in *Birdscapes* (Arlington, Va.: U.S. Fish and Wildlife Service, 2002), 26–27.

7. Christopher Pearl, telephone conversations with author, Corvallis, Ore., February 24, 2004, and July 7, 2004; also, e-mail correspondence with author, March 18, 2004, June 21, 2004, and July 1 and 7, 2004, and letter to author, March 18, 2004.

8. Solicitation no. 00SQ100360, Blasting for Frog Relocation, Bureau of Reclamation-PRNo, 1150 Curtis, Ste. 100, Boise, Idaho 83706-1234 (issue date October 4, 2000), C-1.

9. Gerry Dilley, telephone conversation with author, Yakima, Wash., July 8, 2004.

10. Pearl, telephone conversation.

11. John Faber, telephone conversation with author, Yakima, Wash., July 8, 2004.

12. FitzPatrick, e-mail correspondence.

13. Robert Caudill, field interview by author, Frenchburg, Ky., August 1992.

14. Richard White, field interview by author, Sudith, Ky., September 1991.

15. Richard Hunter, David Cade, and Billy Osborne, field interview by author, Scranton, Ky., August 2002.

16. Subcommittee on Small Water Storage Projects, *Low Dams: A Manual of Design for Small Water Storage Projects* (Washington, D.C.: U.S. Government Printing Office, 1939), 360–361.

17. John H. Klippart, *The Principles and Practice of Land Drainage* (Cincinnati: Robert Clarke, 1861), 39–40, 177–179.

18. Earl J. Osborne, field interviews by author, Frenchburg, Ky., August 1989.

19. Richard Hunter, field interview by author, February 13, 2006.

20. John D. Smith, field interviews by author, Frenchburg, Ky., August 1989.

21. John D. Smith, telephone conversation with author, Sudith, Ky., July 24, 2006.

14. Highways and Waterways

1. Charles G. Elliott, *Practical Farm Drainage* (New York: Wiley, 1903), 84–92.

2. Charles G. Elliott, *Engineering for Land Drainage: A Manual for the Reclamation of Lands Injured by Water,* 3rd edition (London: Wiley, 1919), 273, 339–340.

3. Quincy A. Ayres and Daniel Scoates, *Land Drainage and Reclamation* (New York: McGraw-Hill, 1928), 130.

4. M. D. Adam and M. J. Lacki, "Factors Affecting Amphibian Use of Road-Rut Ponds in Daniel Boone National Forest," *Transactions of the Kentucky Academy of Science* 54 (1993): 13–16.

5. James Kiser, telephone conversation and e-mail correspondence with author, Whitley City, Ky., December 21, 2004.

6. Doreen Miller, telephone conversation with author, Franklin, Ky., April 7, 2004; also, field interviews by author, May 20, 2004, and October 27–29, 2004, and e-mail correspondence with author, December 20, 2004.

7. Arthur C. Parola Jr., telephone conversation with author, January 4, 2006, and July 7, 2006; also, field interview by author, June 26, 2006, and February 21–22, 2006, and e-mail correspondence with author, March

23, 2005, and July 7, 2006.

8. Matthew Thomas and Art Parola, field interview by author, Morehead, Ky., May 3, 2006.

9. Brooks M. Burr and Melvin L. Warren Jr., *A Distributional Atlas of Kentucky Fishes,* Scientific and Technical Series No. 4 (Frankfort: Kentucky Nature Preserves Commission, 1986), 80.

15. Do Waders Come in Size 4?

1. Loretta Roach, field interview by author, Dewitt, Ky., October 18, 2003.

2. Loretta Roach, telephone conversations with author, Dewitt, Ky., February 16, 2005, and March 31, 2006.

3. Phyllis Allison, field interview by author, Prestonsburg, Ky., February 15, 2005.

4. Beverly McDavid, e-mail correspondence with author, Sandy Hook, Ky., April 6, 2006.

5. Margaret Golden, e-mail correspondence with author, Lexington, Ky., April 7 and 9, 2006.

6. McDavid, e-mail correspondence.

7. Ronetta Brown, e-mail correspondence with author, Morehead, Ky., September 27, 2005, and April 7, 2006.

8. Golden, e-mail correspondence.

9. Jim Enoch, telephone conversation with author, Shawnee, Okla., June 3, 2003.

10. Evelyn Morgan, field interview by author, Morehead, Ky., March 20, 2001.

11. Todd Watts, field interview by author, Morehead, Ky., March 21, 2001, and May 12, 2001.

16. Flourishing Fauna

1. A. E. Plocher and J. W. Matthews, *Assessment of Created Wetland Performance in Illinois,* Natural History Survey Special Publication No. 27 (Champaign: Illinois Department of Natural Resources, 2004), 2, 3.

2. U.S. Army Corps of Engineers, *Wetland Planting Manual,* CENAB-OP-RPA (Carlisle, Penn.: Carlisle Regulatory Field Office, U. S. Army Corps of Engineers, 2003).

3. April D. Haight, "Evaluation of the Success of Constructed Wetlands in the Cave Run Lake Watershed," Master of Science thesis, Morehead State University, Morehead, Kentucky, 1996.

4. Jim Lemke, field visit with author, Lexington, Ky., September 8, 2003; also, e-mail correspondence, January 15, 2004, and February 9, 2004.

5. Paul L. Errington, *Muskrats and Marsh Management* (Lincoln: University of Nebraska Press, 1961), 80–81.

6. Ibid., 39, 80–81.

7. Terry Moyer, field interview by author, West Brooklyn, Ill., May 23 and 24, 2005; also, telephone conversations with author, May 25, 2005, and July 18, 2005.

8. Dennis Eger, telephone conversation with author, Montgomery, Ind., October 23, 2006; also, e-mail cor-

respondence with author, June 24 and 25, 2004, October 8 and 12, 2004, and December 1 and 10, 2004.

17. Fixing Failed Wetlands

1. A. E. Plocher and J. W. Matthews, *Assessment of Created Wetland Performance in Illinois,* Natural History Survey Special Publication No. 27 (Champaign: Illinois Department of Natural Resources, 2004).

2. Ibid., 2.

3. William J. Mitsch and J. G. Gooselink, *Wetlands,* 2nd edition (New York: Van Nostrand Reinhold, 1993), 592.

4. Billy Osborne, field interview by author, Pomeroyton, Ky., July 8, 2004. Tammy Cox, telephone conversation with author, Pomeroyton, Ky., July 8, 2004.

5. Henry F. French, *Farm Drainage* (New York: Orange Judd, 1903), 169.

6. Manly Miles, *Land Draining: A Handbook for Farmers on the Principles and Practice of Farm Draining* (New York: Orange Judd, 1893), 106.

7. Dennis Eger, telephone conversation and e-mail correspondence with author, Montgomery, Ind., October 21, 2004; also, field interview by author, June 6–8, 2005.

8. Pat Hahs, telephone conversation with author, Almo, Ky., December 15, 2004. Charlie Wilkins, telephone conversation with author, Hickory, Ky., December 21, 2004.

9. Albert Surmont, telephone conversation with author, Farmers, Ky., November 18, 2004.

10. Mark Lindflott, field interview by author, Des Moines, Iowa, March 9, 2004; also, telephone conversation with author, April 20, 2004.

11. Alice L. Thompson and Charles S. Luthin, *Wetland Restoration Handbook for Wisconsin Landowners,* 2nd edition (Madison: Bureau of Integrated Science Services, Wisconsin Department of Natural Resources, 2004), 105.

12. John Meredith and George McClure, field interview by author, Jackson, Ohio, January 8, 2005.

13. Richard Hunter and Dale Darrell, field interview by author, Salt Lick, Ky., July 28, 2003.

14. Thad Ross and Amy Ross, field interview by author, Leesburg, Ky., July 12, 2003.

15. Anthony Utterback and Sonny Utterback, field interview by author, Cranston, Ky., July 24, 2004.

16. Dennis Eger, telephone conversation and e-mail correspondence with author, Montgomery, Ind., June 18, 2004, and October 23, 2006.

17. John D. Smith, field interview by author, Sudith, Ky., July 1999.

18. Ewell Vice, field interview by author, Flemingsburg, Ky., January 5, 2005.

19. Paul A. Delcourt, Hazel R. Delcourt, Cecil R. Ison, William E. Sharp, and A. Gwynn Henderson, *Forests, Forest Fires, and Their Makers: The Story of Cliff Palace Pond,* Educational Series No. 4 (Lexington: Kentucky Heritage Council, 1999); Paul A. Delcourt, Hazel R. Delcourt, Cecil R. Ison, William E. Sharp, and Kristen J. Gremillion, "Prehistoric Human Use of Fire, the Eastern Agricultural Complex, and Appalachian Oak: Chestnut Forest's Paleoecology of Cliff Palace Pond, Kentucky," *American Antiquity* 63 (1998): 263–278.

20. William Rivers, "Vegetation Record from Bear Pond, Kentucky," unpublished report prepared for the Forest Service, U.S. Department of Agriculture, Daniel Boone National Forest, Kentucky (2000).

18. Finding Funding

1. Scott Manley, field interview by author, Paducah, Ky., March 20, 2003. Jennifer Johnson, telephone conversation with author, Somerset, Ky., March 27, 2003.

2. *Answering the Call: Ducks Unlimited Annual Report* (Memphis: Ducks Unlimited, 2004), vol. 68, no. 7.

3. Loren Renz and Steven Lawrence, *Foundation Growth and Giving Estimates, 2003 Preview,* Foundation Center, April 5, 2004, at http://www.fdcenter.org (retrieved December 20, 2004).

4. Evelyn Morgan, field interview by author, Morehead, Ky., September 22 and 25, 2003.

5. Tony Burnett, telephone conversation and e-mail correspondence with author, Grayson, Ky., December 10 and 20, 2004, and January 24, 2005.

6. *Farm Bill 2002: Resource Conservation and Development Program, Key Points and Program Description* (Washington, D.C.: U.S. Department of Agriculture, 2004).

Glossary

Backhoe: A motorized piece of heavy equipment that features a bucket for digging ditches and a scoop for moving and leveling soil

Bedding: The practice of plowing wet soils in a series of narrow ridges that rise above water and saturated soils; trees and crops planted on the bedded soils have an improved chance of survival

Biological evaluation: A report prepared by the U.S. Forest Service that documents possible effects of implementing a proposed action on federally listed endangered and threatened species

Bog: A highly acid wetland that consists mainly of sphagnum mosses in an abundance of peat; a bog has wet, spongy ground that is too soft to bear the weight of any heavy body on its surface

Borrow pit: An area where soil has been removed for constructing improvements such as a road, parking lot, or dam

Brush ditch: A covered ditch that contains shrubs or trees in the bottom to assist in carrying water beneath the surface

Clay tile: A section of pipe constructed of fired clay, designed to carry water beneath the ground; generally round in shape, clay tiles are from 1 to over 12 inches in diameter

Core: A compacted area of soil beneath a constructed dam, designed to prevent water from leaking under the dam; the core is generally constructed of fine-textured soil, such as clay or silt loam

Ditch: A trench dug in soil for draining surface water and groundwater

Dozer: A motorized piece of heavy equipment on metal tracks; its large blade is used to move and shape soil

Dragline: An old-fashioned motorized machine; it uses cables and chains to pull a large bucket to dig deep ditches

Dredge: A motorized machine designed to work in water; it removes mud for the purpose of digging ditches

Emergent wetland: A shallow-water wetland that contains plants emerging from the water such as cattails and bulrushes; during the growing season, emergent wetland often contains saturated soils and water

Ephemeral wetland: A wetland whose surface water dries annually or in drought years; fish are generally absent from an ephemeral wetland

Excavator: A large motorized heavy-equipment machine on metal tracks; it features a large bucket on the end of a long boom and is used to rapidly dig ditches and holes

Facultative wetland plant: An aquatic plant that grows both in wetlands and on drier ground; willow, sweet gum, joe-pye weed, and cardinal flower are examples of facultative plants

Fen: A wetland or marsh that collects nutrients and is covered with sedges and aquatic grasses; a fen contains an accumulation of peat (consisting mainly of sedges) and is often alkaline

Forested wetland: A seasonally flooded area, often called a swamp, that contains trees such as beech, cypress, elm, maple, pin oak, and willow

Gley: Gray-colored soils that are shown in the "gleyed" color pages in the Munsell Soil Chart; gleyed soils are typically saturated and low in ferric iron (Fe^{3+})

Groundwater wetland: A wetland constructed to expose an elevated water table, much like that seen in a hand-dug well

Hardpan: A buried layer of impermeable soil that prevents water from soaking deep into the ground

Hickenbottom: A vertical pipe that extends above the ground; it aerates a buried drainage system and provides for the rapid flow of surface water into a drainage system

Hillside wetland: A wetland that occurs on sloped ground; hillside wetlands can have saturated soils and standing water during the growing season

Hydric plants: Plants that grow in saturated soil or in standing water and include aquatic species such as sedges, bulrushes, cattails, pondweeds, and water lilies

Hydric soil: A soil that formed under conditions of saturation or flooding, or it may be under water long enough during the growing season to develop anaerobic conditions in the upper part; hydric soils are generally poorly drained with a water table at the surface, they may have water standing over them, and they may be flooded during all or part of the growing season

Hydrologic regime: The duration and frequency that water stands in a wetland and its associated soils

Lands: A series of elevated parallel ridges formed by directionally plowing soils in a field and then planting with crops; the practice of creating lands reduces plant mortality by creating narrow zones of soil that are less likely to be saturated during the growing season

Marsh: An emergent wetland containing shallow water in a pool with grass-like plants such as cattails, bulrushes, and sedges

Mole plow: A piece of equipment designed to cut deep channels for moving water below the surface of the ground; it was first pulled behind livestock in the late 1700s and later was motorized; its primary use has been to drain wetlands for agriculture

Moor: A tract of wasteland sometimes covered with heath, often elevated and marshy, abounding in peat

Morass: A tract of low-lying soft, wet ground that may be swamp, marsh, or bog

Obligate wetland plant: A plant adapted to growing in water and saturated soils and whose presence usually indicates a wetland; water lilies, cattails, and buttonbush are examples of obligate plants

Open ditch: A trench that is not covered with soil

Oxbow wetland: A wetland formed from an abandoned river or stream channel that often has a long, narrow, and curved shape

Plastic drain line: A long, continuous pipe made of plastic and containing narrow slits that is buried in the ground for removing surface water and groundwater from wetlands

Pocosin: Meaning "swamp-on-a-hill" in the Algonquin Indian language, a type of bog found throughout the Atlantic Coastal Plain from southeastern Virginia to northern Florida, and west to Mississippi; a high water table, a large quantity of sphagnum moss, and the slow decay of dead plants contribute to the deep peat and acidic soils of these forested and shrub wetlands

Pothole: An emergent wetland that is generally isolated from rivers and streams

Redoximorphic feature: Mottles of ferric iron (Fe^{3+}) in soil are one type of redoximorphic feature that appear as red blotches; redoximorphic features are generally caused by oxygen entering seasonally saturated soil by plowing, or along root channels and animal burrows

Riffle: A shallow area in a stream where water flows rapidly over a gravel or rocky bed

Riparian: The area of low-lying land that is found along rivers and streams; the riparian area is usually subject to flooding

Rod: A unit of measure equal to 5.5 yards, or 16.5 feet

Sarcle: To weed, or clear of weeds, with a hoe

Scoop, mule: A metal shovel, 3 to 4 feet wide, attached to two handles, and pulled by one or two mules or horses for moving soil

Scraper: A motorized piece of heavy equipment that operates on rubber tires and is used for moving large quantities of soil

Scrub-shrub wetland: A wetland dominated by shrubs such as alder and willow

Seasonal wetland: A wetland that contains water after heavy rains; it generally dries one or more times during the year

Silt basin: A constructed depression or well designed to trap and remove suspended soil from a buried drain line in order to keep it from plugging; silt basins were often constructed at drainage line junctions or where drainage lines changed from a steep to a gradual slope

Slough: A wetland containing deep mud, a bog, or quagmire is a depression in the prairie that is often dry; the term may also be used to describe a water channel or pond filled by freshets

Spillway: A constructed path that allows overflow water to leave a wetland by traveling around the dam; the spillway is usually located on undisturbed ground, is lower than the dam, and has a gradual slope to prevent erosion

Spring: A wet area where water emerges under pressure from the ground

Surface inlet: A constructed pathway by which air and water are designed to enter a buried drainage system; surface inlets are constructed of rock, gravel, clay tile, metal, and plastic and are often covered with soil so as to not affect the cultivation of crops over the top of them; their presence often indicates the historic presence of a wetland

Surface-water wetland: A wetland constructed to hold rainfall and runoff by preventing water from flowing downhill or from soaking into the ground

Swamp: A tract of low, spongy land so saturated with water as to be unfit for tillage; the term was also used to describe a wetland containing trees and shrubs

Track-hoe: *See* Excavator

Water table: The underground surface below which the ground is saturated with water; this zone of saturation is at its highest average depth during the wettest season

Waterway ditch: A shallow, open trench with gradual slopes designed to resist erosion and which often contains a buried drain line; grassy waterway ditches are generally firm enough to be crossed with a tractor

W ditch: A pattern of shallow, parallel ditches that is formed when a wetland is plowed in a manner to create higher ridges for planting (*see also* Lands)

Wetland: An area that contains shallow water during all or part of the year and soils that are saturated during all or part of the year; a wetland contains plants that are adapted to living in the low-oxygen conditions of water and saturated soils

Wet meadow: A wetland containing saturated soils and little, if any, standing water that is dominated by aquatic grasses, sedges, and rushes

About the Author

Tom Biebighauser graduated from the University of Minnesota with a B.S. in wildlife biology in 1978 and began a career with the Forest Service as wildlife biologist at the Superior National Forest in Minnesota, moving on to the Daniel Boone National Forest in Kentucky in 1988. He began restoring wetlands in 1982 in northeastern Minnesota and has since established over 950 such sites in Minnesota, Kentucky, Ohio, and British Columbia. Tom has successfully written hundreds of funding proposals and has taken the lead in completing numerous partnership projects for restoring emergent, ephemeral, forested, and wet meadow wetlands on public and private lands. In 2003, he wrote and published the book *A Guide to Creating Vernal Ponds* in cooperation with Ducks Unlimited, Inc. and the Izaak Walton League of America, distributing over 31,000 copies to date. He finds it rewarding to repair wetlands that have failed and to assist land managers who are working to begin wetland restoration programs. Tom enjoys leading hands-on workshops where participants learn about wetland restoration by actually constructing a wetland; he has taught these workshops in Alabama, Illinois, Kentucky, Maryland, Minnesota, New York, North Carolina, South Carolina, and British Columbia. His latest program involves helping teachers build wetlands at schools for use as outdoor classrooms. He encourages hundreds of students each year to put on waders and venture into wetlands for exploration. For six years, he has served as chairperson for the Cave Run Chapter of Ducks Unlimited. Governor Brereton Jones commissioned Tom a Kentucky Colonel in 1995. In 1999, he received an award from Goodyear and the National Association of Conservation Districts for outstanding accomplishments in resource conservation practices. His passion for wetland restoration has been recognized by three National Taking Wing Awards: for leadership, community involvement, and accomplishments. In 2005, the Rowan County Conservation District presented Tom with its Honor Award. He received the Vocation Award from the Morehead Rotary Club the same year. Congressman Hal Rogers presented Tom with the Kentucky PRIDE Award in May 2005, for leadership in wetland restoration.

Index

ditch(es) *(continued)*
    W, 26–27
    waterway, 22, 24
dragline, 20, 22, 86–87
drainage
    benefits of, 3–8, 11–14, 44, 74–75, 195–196
    Elkington method of, 34–35, 57
    emergent wetlands, 25, 55, 65, 66, 69, 77, 81, 98, 196
    ephemeral wetlands, 4, 7, 25, 34, 60, 62, 64–65, 80, 82–83, 99
    forested wetlands, 4, 60, 65, 66, 68, 71–73, 99
    in forestry, 14–15, 27
    hand-dug, 4–7, 9, 17, 63, 66, 69, 71
    by machine, 31, 47–57
    scrub-shrub wetlands, 25, 69, 99
    test for, 4
    and wells, 35, 57
    for wetland construction, 120, 168–170
    wet-meadow wetlands, 4, 7, 25, 60, 78–79, 99
drainage districts, 8, 96
drain lines
    of brush, 3–4, 31
    of clay tiles, 5–6, 39–57
    depth of, 25, 43, 45, 74, 86, 104, 129, 197, 206–210
    lifespan of, 3–6, 45, 72, 75, 86
    locating in ground, 44–46, 76, 92, 94, 104–105, 209
    of Orange-Burg pipe, 73
    outlets, 43, 46, 48, 58, 67, 75, 104, 208
    placement of, 40–41, 53
    of plastic, 31, 49–53
    of poles, 31–33
    of rock and stone, 7, 31, 34–38, 57
    slabs, 7, 31, 33, 39, 71
    of wood boxes, 6–7, 30–31, 71–72
drained wetland, identification of, 17, 36–37, 54–57, 62–99, 100–101, 107–108, 138
dredge(s), 17, 21
Ducks Unlimited, Inc., 111, 128, 135, 160, 203, 217–219
dump truck(s), capacities of, 189
dynamite, 17, 21, 23, 163

Eger, Dennis, 141–143, 198, 207–208, 214–215
electromagnetic induction meter (EM 38), 68, 72–73
Elkington drainage method, 34–35, 57
endangered species
    bog turtle, 89–90
    Cumberland elktoe mussel, 117
    Indiana bat, 116–117, 139, 144, 153, 172
    regulations, 116–117
    Virginia big-eared bat, 153
explosives
    for wetland construction, 158–164
    for wetland drainage, 21, 23

Faber, John, 164
farm ponds
    construction from wetlands, 59–61, 78
    returning to wetlands, 167–168, 211

Feaster, Brad, 135–136
fen, 4
fill (soil)
    construction of wetlands on top of, 155–156, 183–185, 187, 213
    placement in wetlands, 85, 90, 95, 102, 112, 160, 181
filling
    of wetlands by hand, 5, 7, 18, 21
    of wetlands by machine, 44, 58–61, 65–67, 69, 77–78, 80–81, 167
fire plow, 15–16
FitzPatrick, Denny, 158, 160, 164
flashboard riser, 171, 200, 208
Flegel, Kathy, 122, 127–129, 134
Florida, 29, 108, 111–112
Flowers, James, 41–44, 92, 98
frogs
    and drainage, 8–9, 66, 79, 91, 108
    Oregon spotted, 161–164
    use following wetland restoration, 134, 144, 153, 156, 158, 160, 164, 184, 193, 210, 213
funding, sources and strategies, 217–221

Georgia, 9, 14
Golden, Margaret, 185–186
gravel
    buried layers and wetland construction, 118–127
    pits, 155–158, 164
Great Britain, 5
Great Dismal Swamp, 4
Great Smoky Mountains National Park, 106
ground-penetrating radar (GPR), 68, 72, 206
gum ponds, 138–141

Hahs, Pat, 208
Hakala, Jon, 158, 160
Hickenbottom surface inlet, 54–56, 58, 81–82, 91
Hunter, Richard, 145–146, 152, 166, 168, 175, 211
Hurst, Don, 9, 21, 84, 104

Illinois, 197
Indiana, 8–9, 118, 135–136, 141, 143, 198, 207–208, 214–215
Iowa, 40, 55, 104, 208, 210
Ireland, 11, 34

Jackson, Lacy, 104
Jackson, Michael, 213
Johnson, Jennifer, 217
Johnston, John, 4–5, 39, 73–76
Johnstone, John, 3–4, 22, 27, 33–34, 55, 65, 90

Kansas, 135
Karr, Mike, 160
Keenon, John, 9
Kentucky counties
    Ballard, 209
    Bath, 13, 19–21, 23–24, 52, 56, 88, 139, 154, 174, 187, 211

success of, 194–195, 199
    and surface water, 4, 102–103, 118–154, 178, 216
    in timber harvest areas, 16, 144–145, 152
wetland values, 1, 182–186
Whitecross, Scott, 82–83
white oak, 79, 152, 193

Wilkins, Charlie, 208–209
Williams, Richard, 213
willow, 9, 18, 33, 39, 60, 77, 80–81, 83, 140, 160, 201
Wisconsin, 55, 210

Yearick, Melissa, 75